Jose Adonias De Franca
Celso Amorim Câmara

# Cyclo-oxidation of nitrogen derivatives of lapachol

**Jose Adonias De Franca**
**Celso Amorim Câmara**

# Cyclo-oxidation of nitrogen derivatives of lapachol

## In search of potentially active derivatives

**Imprint**
Any brand names and product names mentioned in this book are subject to trademark, brand or patent protection and are trademarks or registered trademarks of their respective holders. The use of brand names, product names, common names, trade names, product descriptions etc. even without a particular marking in this work is in no way to be construed to mean that such names may be regarded as unrestricted in respect of trademark and brand protection legislation and could thus be used by anyone.

Cover image: www.ingimage.com

This book is a translation from the original published under ISBN 978-613-9-65601-1.

Publisher:
Sciencia Scripts
is a trademark of
Dodo Books Indian Ocean Ltd. and OmniScriptum S.R.L publishing group

120 High Road, East Finchley, London, N2 9ED, United Kingdom
Str. Armeneasca 28/1, office 1, Chisinau MD-2012, Republic of Moldova, Europe
Printed at: see last page
**ISBN: 978-620-7-79191-0**

# SUMMARY

# DEDICATORY

To the love of my life Meire Falcao

and all my family.

# ACKNOWLEDGMENTS

To my mother Elizabete Alves dos Santos, who, without her support, would have made my road to scientific life very difficult. To my stepfather Rivaldo Paulino, for his great friendship and help with his friendly advice. To my brothers, my friends, who are always there when I need them. To my uncle Anibal, my kickboxing teacher, who gives me the greatest support to follow my destiny, whether here or abroad. To my father-in-law, Mr. José Bezerra (*in memoriam*), for giving me the tools I needed to carry out some experiments. To my mother-in-law[a] . Iracema, always a friend. To my friends Dilmo Marques and Patricia Ferreira, for their sincere friendship over all these years. To my friend and almost brother Càndido Manuel, the human encyclopedia, whom I already consider to be a great master of chemistry. To my friend Gilson Bezerra, who even though I only knew him for a short time, we soon became friends and discussed many chemistry subjects. To my friend Edmilson Clarindo, who always gave me support and help whenever I needed it. To my friend Litivack, temporary jiu-jitsu teacher and academic road companion. To my friend Thiago Selva, who was always there to help me with my computer problems. To my friend Rômulo, always friendly. To my laboratory friends Augusto, Rogério, Patricia, Thamiris and Wilson. To the technicians at the analytical centers Vicente (UFPB) and Ricardo (UFPE) for the NMR and IR spectra. To UFF professors Vitor Ferreira and Fernando de Carvalho and master's student Priscila, for the $^{1}$H and $^{13}$C spectra. To the professors, Bogdan Doboszewski, who always takes the time to clear up a doubt; Ronaldo Nascimento, Roberto Antunes, Clàudio Câmara (the coffee fan); and Professor Celso de Amorim (a great figure), who has been guiding me since my scientific initiation, a special thanks, I hope one day to reach your level of knowledge. And my biggest thanks go to the girl who is always by my side. Meire Falcâo, my guardian angel and my fiancée.

# EPILOGUE

*"There is no greatness when there is no simplicity."*

*Count Leon Nikolayevich Tolstoy*

# SUMMARY

In the course of the studies on 1,4-naphthoquinones, a methodology for synthesizing nitrogen rings based on palladium-catalyzed allylic oxidation reactions was explored. It was possible to obtain some compounds unpublished in the literature during these studies, and the reaction to obtain nor-lapachol through the condensation of lausone with isobutyraldehyde was significantly improved. A new methodology was developed for the formation of dimeric naphthoquinone molecules, 4 of which are new. The compounds were tested against the human topoisomerase enzyme II and in toxicity tests using *Artemia salina*.

# 1. INTRODUCTION

## 1.1. Quinones and their importance

A range of hydroxynaphthoquinones and their derivatives have molluscicidal (Lima, 2002), antibacterial (Riffel, 2002) and trypanosomicidal (Tonholo, 1998) activity. It is already well known that the quinone nucleus has great potential to act as an antimycobacterial (McKinney, 2000). In general, the most representative natural quinones are of vital importance to higher plants, fungi, lichens, bacteria, algae, echinoderms, viruses and some arthropods. The distribution of these substances in various organisms possibly implies multiple biological functions, acting in their various biochemical cycles (Thomson, 1971). Recent publications have expanded the chemical study of the lapachol molecule with new reactions in the prenyl chain used as a nucleophile, resulting in nitrogen heterocycle rings unpublished in the literature, as well as open-chain nitrogen derivatives, according to Scheme 1 (Camara *et al*, 2001).

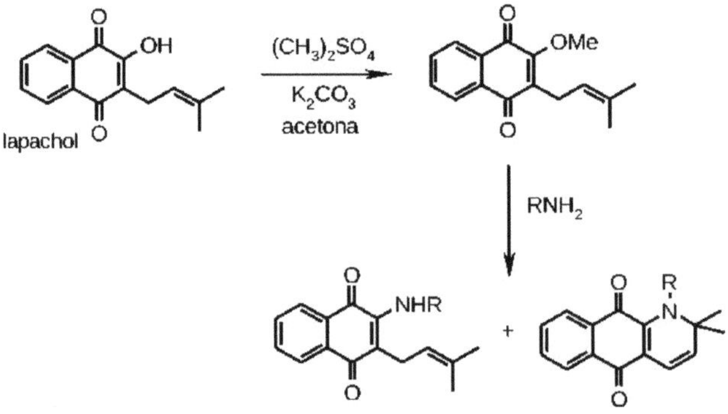

Scheme 1: Nitrogen derivatives of lapachol

Synthetic routes are already known in the literature to arrive at methyl-substituted derivatives in positions 2 and 3, from right to left, as shown in Figure 1, (Behforouz, 1997) (Catti, 2008) of an alkaloid (**1**) that was tested in the 1970s as a possible adjuvant in the treatment of tuberculosis (Vaitiligam, 2004) (Xu, 2004).

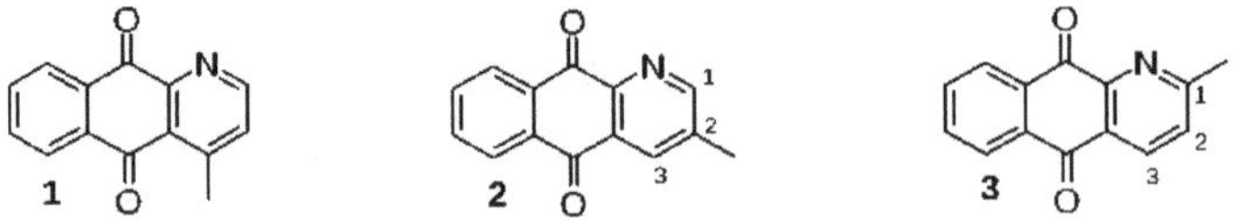

Figure 1: Cleistofolin **1** and its analogues

Our research group has published work on amino-cyclized naphthoquinonic substances with molluscicidal activity against the *Biomphalaria glabrata* snail, a vector for transmitting schistosomiasis, and with cytotoxicity tests against *Artemia salina* Leach (Barbosa, 2005). In order to make the bond between the nitrogen and the carbon in the allylic position, the palladium catalysis process was carried out, since the selective activation of the CH bond in the allylic position, marked next to the R group in Figure 2, is favored by catalysis with transition metals (Wu, 2009), as can be seen in Figure 2.

R=H;R'=CH₃ amino-nor-lapachol
R=CH₃;R'=H amino-iso-norlapachol

R''=H;R'''=CH₃ **3**
R''=CH₃;R'''=H **2**

Figura 2: Proposal for a palladium-catalyzed reaction

## 1.2. Palladium compounds widely used in organic synthesis

Two types of palladium compounds are widely used in organic synthesis: palladium (II) compounds and palladium (0) complexes (Tsuji, 1995). Those with palladium valence (II) are marketed as PdCh and Pd(OAc)2. Palladium chloride is stable, but has low solubility in organic solvents and water. It is soluble in dilute HCl and becomes soluble in organic solvents by forming complexes with various ligands (Tsuji, 1995). Palladium acetate is commercially available, soluble in organic solvents and can be obtained by dissolving metallic palladium in acetic acid containing nitric acid.

It is well known in the literature that allylic functionalization can be carried out via catalysis with palladium compounds. Acetoxylation is a proposed way of activating the allylic position, since if the leaving group is abandoned, a complex is formed with the palladium, favoring the entry of a nucleophile (Larock, 1991), as can be seen in figure 3:

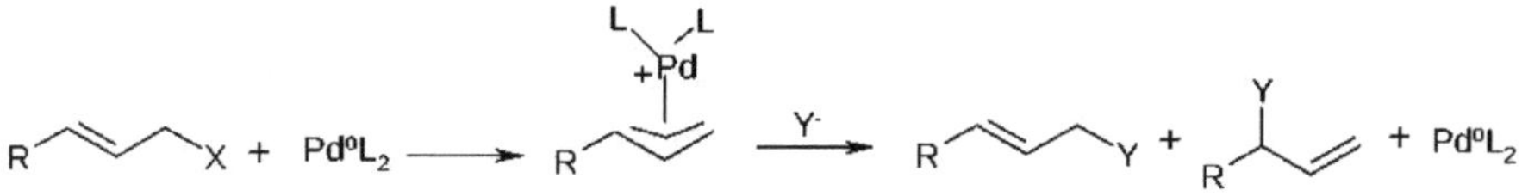

Figura 3: General scheme for the formation of π -allyl palladium complexes

## 1.3. Quinolines and synthetic similarity

A large number of biologically active substances from natural and synthetic sources are in the quinoline and quinoline group (Kolodina, 2007). Studies show that compounds derived from quinolines are used in the treatment of Parkinson's disease (Liu, 2006). Quinoline was first isolated by Runge in 1834 from the distillation of tar and was called "Leukol". It was also obtained in 1842 by Gerhardt by distilling cinchonine and quinine with alkali, then called "Chinolein" (Jones, 1977). Dewar in 1871 suggested that quinoline was related to pyridine, just as naphthalene is to benzene. The structure was confirmed by synthesizing it from 2-nitrocinnamaldehyde, as shown in Scheme 2.

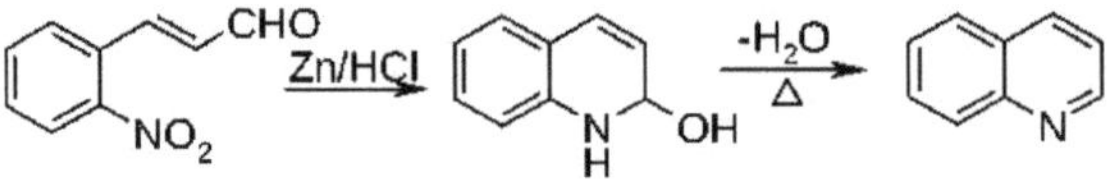

Diagram 2: Confirmation of the quinoline structure

The general procedure by which an aromatic amine is condensed with an α,β-unsaturated compound or its precursor, in the presence of an oxidizing agent, represents the classic method for obtaining quinolines. This method is known as Skraup synthesis (Ghera, 1981) and is shown in Diagram 3.

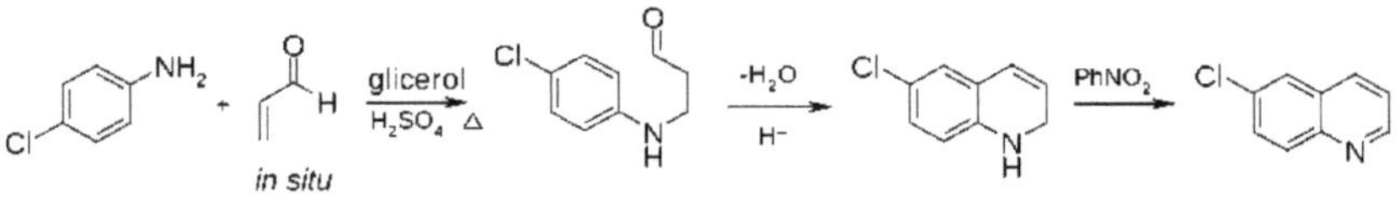

Diagram 3: Classic method for obtaining quinolines

In Skraup's synthesis, glycerol is first dehydrated to acrolein *in situ*. The Michael attack takes place under acid catalysis, generating 1,2-dihydroquinoline after cyclization and dehydration. Nitrobenzene acts as an oxidizing agent, removing the two hydrogens via a radical reaction, re-establishing the aromatization of the system.

Anilines react with 1,3-diketones as well as acetylketones in the presence of acids (Combes Synthesis) to generate enaminones, which by acid cyclization lead to substituted quinolines of type (A), Scheme 4 (Palmer, 1967).

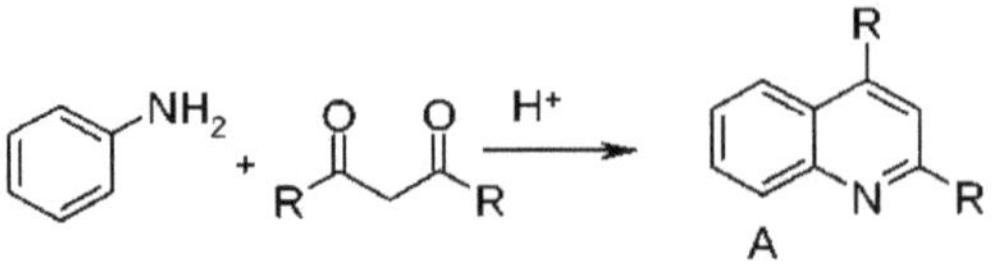

Scheme 4: Formation of substituted quinolines

Observing that there is an immense variety of quinoline syntheses, our research group focused on the synthesis of cleistofoline analogues (**1**), 2-methyl (**3**) and 3-methyl (**2**), from allylic oxidation reactions of naphthoquinone derivatives such as nor-lapachol (**4**) and nor-isolapachol (**5**), using palladium compounds as catalysts. In short, the methodology for cyclizing quinolines can be extended to the synthesis of naphthoquinone analogues, which is the main focus of this work, as can be seen in Scheme 5.

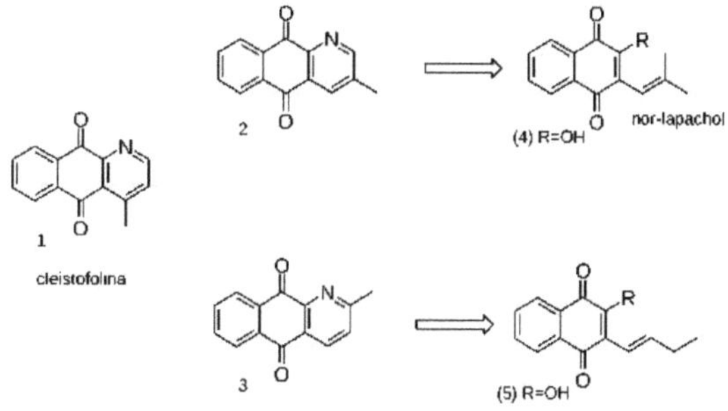

Diagram 5: Retrosynthetic analysis

## 1.4. **Naphthoquinones and tuberculosis**

Tuberculosis is a disease with major social repercussions and historically associated with poor health and hygiene conditions, particularly in developing countries. Its etiological agent is the microorganisms *Mycobacterium tuberculosis, Mycobacterium avium, Mycobacterium bovis,* among others. Currently, an estimated two billion people are infected worldwide (Sriram, 2004). Tuberculosis (TB) is a serious bacterial infection and it is estimated that one third of the world's population is infected with *Mycobacterium spp* (Smieja, 2009). Bovine tuberculosis due to *M. bovis is* the major cause of gastrointestinal tuberculosis in developing countries and is caused by the use of unpasteurized bovine milk (Bonsu, 2000). Around 9 million people develop pulmonary tuberculosis, resulting in 3 million deaths a year from the disease (McKinney, 2000). The drugs that revolutionized the treatment of tuberculosis in the 1970s had their usefulness greatly diminished in the following decade due to the emergence of multiple resistance. This led to a huge search for new drugs capable of dealing with this deadly disease, since there is no selectivity for individuals (Vaitiligam, 2004) (Xu, 2004). Among these new drugs, many compounds of plant origin have been tested, including cleistofolin (**1**) (Peterson, 1992). Cleistofolin (**1**) Scheme 5 is an aza-anthraquinone, or 4-methyl- benzo[*g*]quinoline-5,10-dione, and was initially isolated as a natural product in several species of the Annonaceae family (Waterman, 1985), and among its biological activities described are: antifungal (Hsieh, 2001), cytotoxic against hepatocarcinoma (Hsieh, 2001) and antimycobacterial (Peterson, 1992). Quinones, including naphthoquinones and lapachol, have been shown to be active as antimycobacterials (Tran, 2004). In this work, it is clear that for this activity in particular, 1,4-quinones (alpha-quinones) have generally proved to be more active than isomeric beta-quinones (1,2-quinones). To this end, our work is based on finding an effective method for synthesizing cyclic nitrogen derivatives analogous to cleistofolin (**1**).

### 1.5. **Naphthoquinones and topoisomerase**

DNA supercoiling is a precisely regulated process that influences many aspects of DNA metabolism. Every cell has enzymes whose function is to unwind and/or relax DNA. Enzymes that increase or decrease the degree of DNA unwinding are called topoisomerases, and the DNA property they affect is the linker number (Lk) (Nelson, 2002). These enzymes play an important role in processes such as DNA replication and compaction. There are two classes of topoisomerases. I-type topoisomerases act by transiently breaking one of the DNA strands. The II topoisomerases break both strands of DNA (Nelson, 2002). DNA topoisomerases catalyze topological changes in DNA that are essential for normal cell cycle progression and are therefore preferred targets for the development of anticancer and antimicrobial drugs, for example, ciprofloxacin® which fights *Bacillus anthrax*. Anti-topoisomerase drugs can be divided into two main classes: anti-"cleavable complex" drugs and catalytic inhibitors. Anti-"cleavable complex" drugs are very effective as anti-cancer drugs, but they are also potent inducers of chromosomal aberrations and can cause secondary malignancies. Catalytic inhibitors are cytotoxic, but do not induce chromosomal aberrations. Knowledge of the mechanism of action of topoisomerase inhibitors is important for determining the best anti-topoisomerase combinations with a reduced risk of inducing secondary malignant neoplasms (Loredana, 2000).

It is remarkable that both DNA-helicase and topoisomerase perform their functions without causing permanent alterations to the chemical structure of DNA. DNA-helicase only breaks the hydrogen bonds that hold the two strands of DNA together, without breaking any covalent bonds (Watson, 2006). And although topoisomerases break one or more covalent bonds in the DNA, each broken bond is exactly restored before the DNA is released (Watson, 2006). Instead of altering the chemical structure of the DNA, the action of these enzymes results in a DNA molecule with an altered conformation. It is important to note that these conformational changes are essential, among other activities, for the duplication of the huge double-stranded DNA (dsDNA) molecules that form the basis of prokaryotic bacterial chromosomes. New derivatives of lapachol (**6**), nor-lapachol (**4**) and lausone (**7**), were synthesized and tested against the topoisomerases I and II -α, and the compounds

(**8**, **8a** and **8b**) which showed inhibition of the catalytic activity of the enzyme 11 -α at a dose of 2μM (Cunha, 2006) showed promise (Figure 5). It can therefore be concluded that combating the activity of the topoisomerase enzyme would also combat the proliferation of the viruses that use them (Watson, 2006).

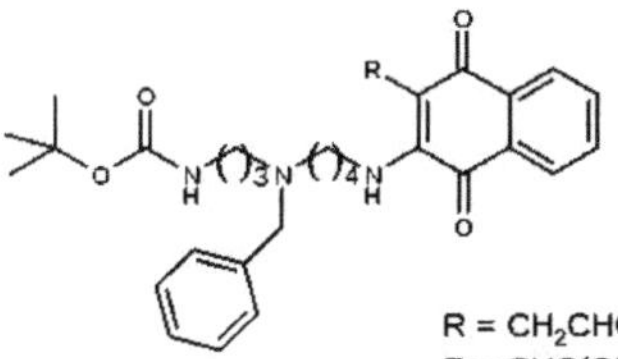

$$R = CH_2CHC(CH_3)_2 \quad \textbf{8}$$
$$R = CHC(CH_3)_2 \quad \textbf{8a}$$
$$R = H \quad \textbf{8b}$$

Figure 4: Promising naphthoquinone derivatives against topoisomerases I and II -α

There is also work related to aminonaphthoquinones and anthraquinones derived from lapachol (**6**) which showed activity against the enzyme topoisomerase II -α (Souza, 2007). This work showed that compounds **9a-d** Figure 6, were significantly active in inhibiting the enzyme at a concentration of 200 μM.

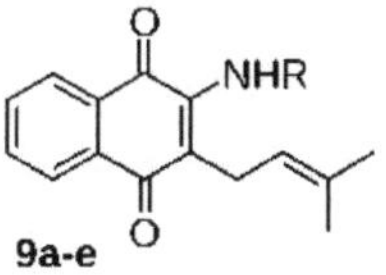

$$R$$
$$a=CH_2CH=CH_2 \quad \textbf{9a}$$
$$b=CH_2C_6H_5 \quad \textbf{9b}$$
$$c=CH_2CH(OCH_3)_2 \quad \textbf{9c}$$
$$d=CH_2CH_2OH \quad \textbf{9d}$$
$$e=CH_2COOH \quad \textbf{9e}$$

Figura 5:  Amino-naphthoquinones that showed activity against topisomerase II -α

## 1.6. Naphthoquinones and leishmaniasis

In addition to biological tests with human topoisomerase II, our research group is interested in testing drugs against leishmaniasis, knowing that it is a parasitic disease caused by flagellated protozoa of the genus *Leishmania,* which occur endemically in 88 countries on four continents, affecting around 12 million individuals (Gontijo & Carvalho, 2003).

Leishmaniases infect a variety of mammal species, including humans, and are transmitted between vertebrate hosts through the bite of female phlebotomines. Leishmaniasis is a rapidly expanding disease and it is estimated that 350 million people are exposed to the risk of transmission, with an annual incidence of around 1.5 to 2 million cases. Approximately 90% of all recorded visceral leishmaniasis (VL) cases come from five countries: Bangladesh, Brazil, India, Nepal and Sudan, while 90% of mucocutaneous leishmaniasis (MCL) cases occur in Bolivia, Brazil and Peru (Gontijo & Carvalho, 2003). In addition to a complex epidemiological chain that makes control actions difficult, different species of the *Leishmania* genus act as etiological agents of the disease. In the Americas, it is accepted that VL is caused only by *L. chagasi,* while for ATL and CML the most frequent etiological agents are *L. braziliensis* and *L. amazonensis* (Gontijo & Carvalho, 2003).

*In vitro* tests showed that the naphthoquinone derivative (**10**) Figure 7, was very active against *L. amazonensis* with its (ICsɑ/24 h= 1.6 0.0 µg/ml), which was compared to the market drug pentamidine isethionate® with (ICsɑ/24 h=0.28 0.05 µg/ml), thus showing promise for the treatment of *L. amazonensis* (Lima, 2004).

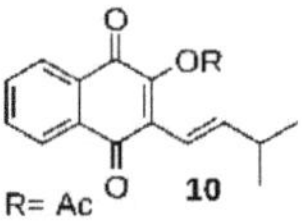

R= Ac   **10**

acetilisolapachol

Figura 6:   Naphthoquinone that showed activity against *L. amazonensis*.

# 2.  OBJECTIVES

Obtaining semi-synthetic derivatives from abundant natural products and their pharmacological evaluation, we intend to obtain the aza-anthraquinones (**2**) and (**3**), (Camara, 2003), structurally analogous to cleistofolin (**1**), using nor-lapachol (**4**) (Barbosa, 2004) derived from lapachol (**6**), and 2-amino-3-(2-butenyl)-1,4-naphthoquinone (**5a**) derived from lausone (**7**) (Hayashi, 1987) (Nagabhushana, 2001). These compounds are abundant raw materials (Camara, 2001) (Camara, 2002), of plant and synthetic origin (Diagram 6). There are already methodologies in the literature based on Diels-Alder reactions to obtain anthraquinones (**2**) (Catti, 2008) and (**3**) (Behforouz, 1997).

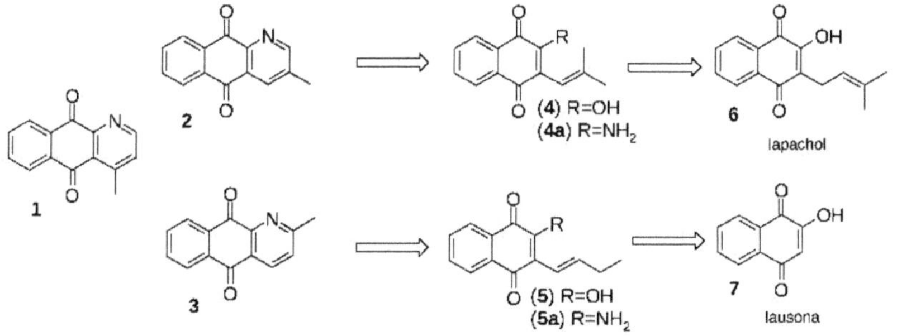

Diagram 6: Synthesis strategy

The aim is to synthesize derivatives (**2**) and (**3**) without using Diels-Alder reactions, which are already well known in the literature. In addition, a new methodological process consists of starting from hydroxy-naphthoquinones by transforming the hydroxyl group of (**4**) and (**5**) into the respective amino group of (**4a**) **and** (**5a**). This involves the methodology developed previously (Camara, 2002), by alkylating the hydroxyl with dimethyl sulphate in a basic medium followed by nucleonic substitution of the 2-methoxyl group in the presence of ammonium hydroxide.

## 2.1.  Methodology

The allylic cyclo-oxidation of nor-lapachol (**4**) and 2-hydroxy-3-(2-butenyl)-1,4-naphthoquinone (**5**) using 3,4-dicyano-5,6-dichloro-benzoquinone (DDQ) is the key step in the functionalization of the side chain (Dudley, 1969), as exemplified for nor-lapachol (**4**) (Scheme 7).

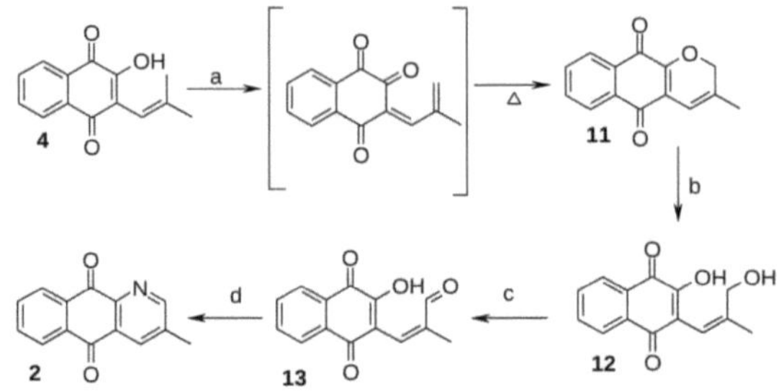

Conditions and reagents: a) DDQ, hexane; b) NaOH aq; c) THF reflux, $MnO_2$ ; d) $NH_3$ aq

Diagram 7: Synthetic proposal

This functionalization can alternatively be carried out by allylic functionalization with (NBS) (Kar, 2003) producing, after intramolecular cyclization *in situ,* the same unsaturated pyran derivative (**11**). This pyran ring, in turn, can be cleaved by dilute base under reflux, since chemically it behaves like a vinyl ester, due to the conjugation between the ether function and the carbonyl in position 4 (Hooker, 1936). The open-chain quinone (**12**) can be oxidized at the allylic hydroxyl under mild conditions using manganese dioxide in THF at reflux (Kirchlechner, 1994), leading to the derivative (**13**). The use of ammonia under mild reflux conditions will lead to the formation of the final aromatic derivative (**2**) 3-methyl-benzo[ *g*]quinoline-5,10-dione analogous to cleistofoline (**1**).

Preliminary investigations have indicated that the conversion of nor-lapachol (**4**) to the pyran ring of (**11**) occurs in exiguous yields of approximately 16 %[2] , as shown in (Scheme 8).

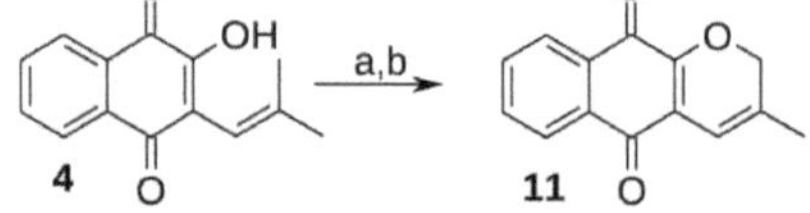

Diagram 8: Formation of vinyl ester

In view of this difficulty, we propose transforming the hydroxyl group of (**4**) and (**5**) into the amino function of (**4a**) and (**5a**). In this way, the cyclization step will lead directly not to pyran derivatives, but to piperidine derivatives, with the correct nitrogen functionalization (Scheme 9).

---

2 Data published at the event held at UFRPE entitled: Jornada de Ensino Pesquisa e Extensâo (JEPEX) in 2006.

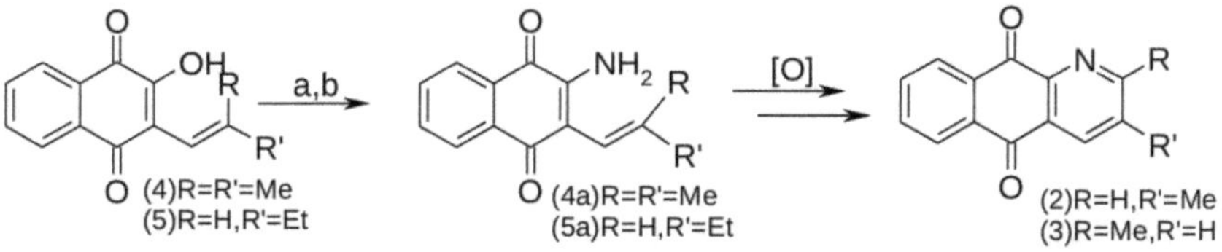

Conditions and reagents: a) Me2SO4, K2CO3, acetone; b) NH4OH, MeOH, reflux

Diagram 9: Alternative route

The amines formed in the previous reaction (**4a**) and (**5a**) are the precursors, *seco-an analogues, of* the respective cyclic compounds (**2**) and (**3**), 2-methyl and 3-methyl analogues of Cleistofoline (**1**) (4-methyl-1-aza-anthraquinone). Preliminary attempts at allylic oxidation carried out by our research group identified selenium dioxide (Hudlicky, 1990) as a promising oxidizing agent among several oxidizing agents capable of functionalizing the allylic position of (**4a**) and (**5a**). Other oxidizing agents tested include DDQ (3,4-dichloro-5,6-dicyano-p-benzoquinone), mercury (II) acetate and manganese dioxide (Diagram 10).

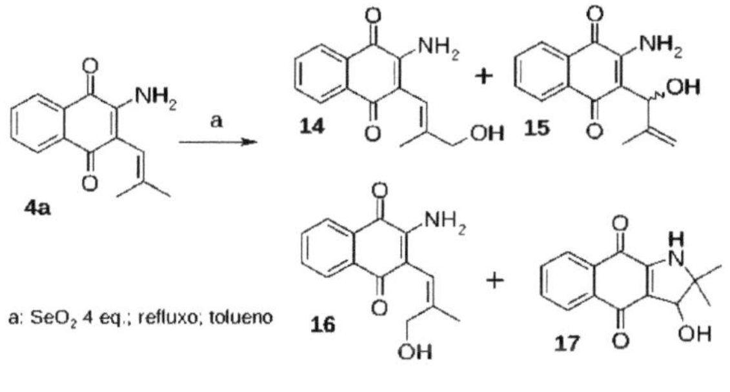

Diagram 10: Possible products of the allylic oxidation of compound 4a

In these tests, the starting compound, the amine (**4a**), was completely consumed after 4 days at reflux in excess SeO2 in the presence of the solvent toluene under reflux. Among the products isolated in the reaction was a polar mixture of several compounds, which suggests that it is a mixture of possible allylic alcohols (**14**, **15** and **16**) resulting from the oxidation of (**4a**) in the allylic position (other possible products would be the respective aldehydes and ketones and/or acids, resulting from the subsequent oxidation of the allylic hydroxyls). A compound of lower polarity in low yield (~10%, p.f. 150-5 °C) was isolated and identified as 3-hydroxy-2,2-dimethyl-2,3-dihydro-1H-benzo[ *f*]indol- 4,9-dione (**17**) by NMR of [1] H and [13] C (CDCl3, 300 MHz, p. 64 and 65) respectively.

The nucleophilic power of the amine group *per se does* not seem to be enough to carry out the desired transformation, i.e. the *in situ* cyclization of the oxidized *seco-amino* derivative (exemplified by **18**) into the respective cyclic analogue (**2**) (Scheme 11).

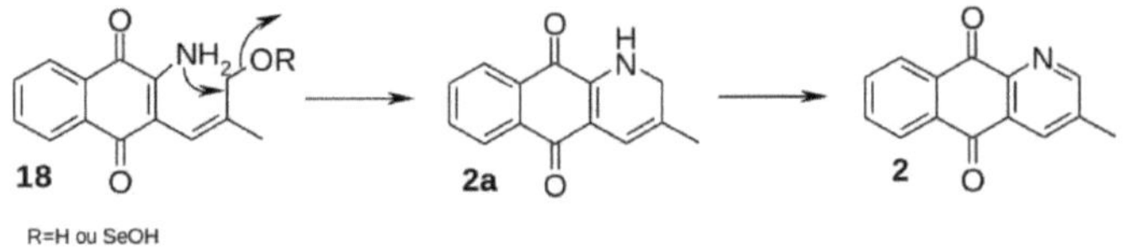

Diagram 11: Nucleophilicity of the amine group

Under the reaction conditions used, the intermediate (**2a**) can be oxidized to the final product (**2**), which is aromatic, due to the excess of oxidizing agent used. Nucleonic substitution will probably still take place in the oxidation intermediate in the form of a selenite ester (formed by the attack of selenium dioxide on the double bond in the allylic position) and, because it is an allylic intermediate, it has increased reactivity. The probable intermediates (**14**) and (**16**) can undergo isomerization of the double bond and/or conjugate addition resulting in the same expected product (Scheme 12).

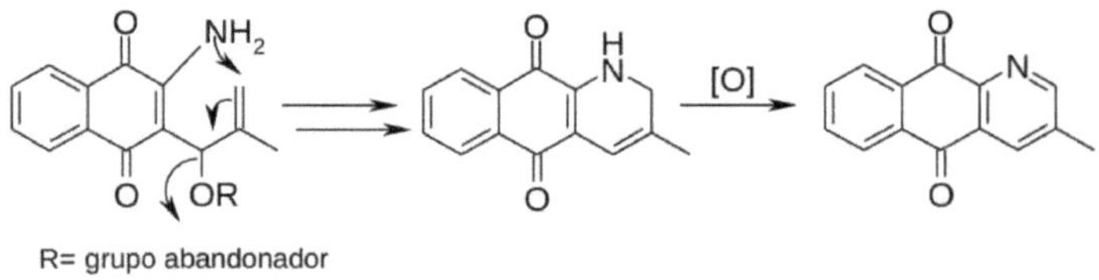

Diagram 12: SN2' substitution

One of the reasons that may explain the low reactivity of the amine group in carrying out the expected intramolecular nucleonic addition may be related to the fact that the nitrogen in these compounds is conjugated to the carbonyl of the quinone nucleus through the double bond, resulting in a vinyl amide that is not very nucleonic (Figure 8).

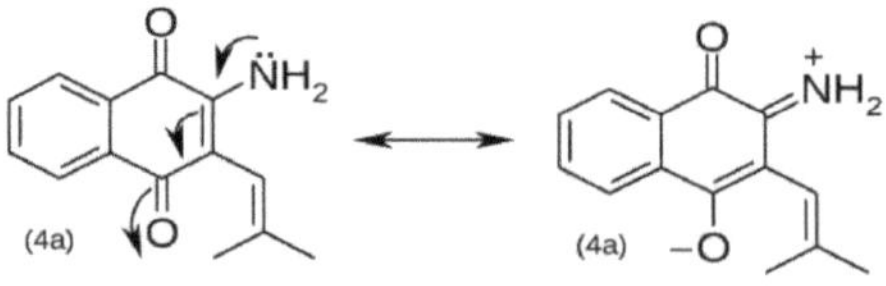

Figure 7: Conjugation of the amine group

Strategies for the cyclization of amines involving the addition of allylic alcohols to amines in heterocyclization reactions are known in the literature. In an initial approach (Larock, 1991) the synthesis of 2,3 and 4-alkyl-quinolines was carried out from the palladium-catalyzed cyclocondensation of 2-iodo-aniline with appropriately functionalized allylic alcohols, in yields ranging from 23 to 62 % (Scheme 13).

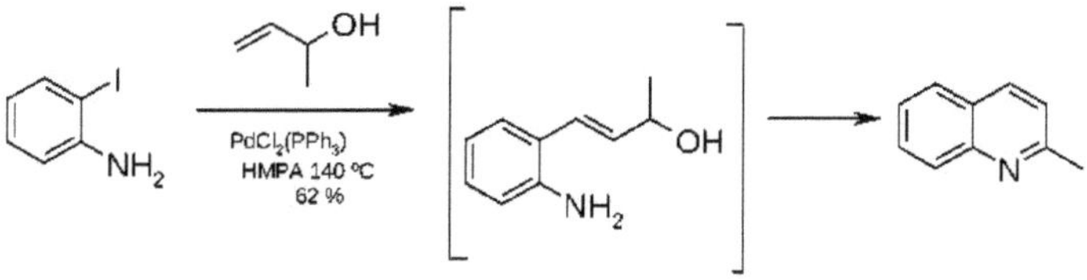

Scheme 13: Cyclization strategy using allyl alcohol

The formation of an arylalkene through the Heck reaction with the iodine derivative in the formation of the C-C bond precedes the intramolecular cyclization reaction, probably involving a step in the formation of a π-allyl-palladium intermediate. The high temperatures used in the reaction favor the formation of the cyclization product.

The allylic oxidation reaction of alkenes using catalytic palladium is also a synthetic possibility to be investigated in the course of this work (Hansson, 1990) (Scheme 14). The allylic acetate thus obtained (or a mixture of isomers) can be taken to a subsequent stage without the need for isolation for the allylic amination reaction under appropriate conditions (Scheme 14).

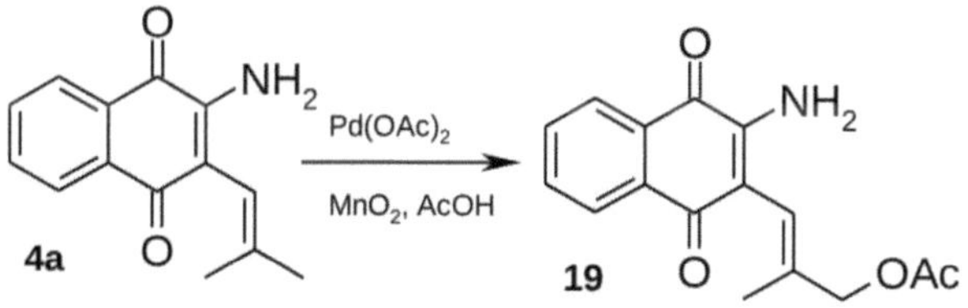

Scheme 14: Allylic acetoxylation of compound 4a

In addition to the nor-lapachol derivatives, which will lead to the 3-methyl cleistofoline analog (2), similar reactions will be studied with *2-hydroxy-3-E-butenyl-1,4-naphthoquinone* (5), which will lead to the synthesis of the 2-methyl cleistofoline analog (3). To do this, it will be obtained by condensing lausone with n-butyraldehyde under conditions already established in the literature (Hooker, 1936). The hydroxyl of (5) will be transformed into the amine (5a) using a procedure already established in the literature

17

(Camara, 2002).

## 2.2 Material and equipment

The infrared (IR) spectra were obtained on a BOMEM MB - Series 100 FT-IR spectrophotometer at the LTF Pharmaceutical Technology Laboratory (UFPB) and the UFPE analytical center, using potassium bromide tablets. The absorption values are given in wave numbers, using the reciprocal centimeter ($cm^1$ ) as the unit. The NMR spectra were carried out at the Institute of Chemistry of the Fluminense Federal University (UFF), the LTF (UFPB) and the Department of Fundamental Chemistry (DQF) of the Federal University of Pernambuco (UFPE).

The products were purified by column adsorption chromatography using silica gel 60 (230-400 Mesh ASTM - Merck) as the stationary phase. The products were purified by preparative thin layer chromatography (SDLC) using silica gel 60 (PF254 with gypsum - Merck) on manually prepared glass chromatoplates (20 x 20 cm).

Chromatography plates (20 x 20 cm) made of silica gel 60 (F254- Merck) were used for analytical thin layer chromatography (ADLC) analysis. The substances were visualized in CCDA with the naked eye, with a UV lamp and by immersion in an alcoholic solution of ferric chloride (specific enolic hydroxyl developer).

The solvents were used as purchased, without purification or distillation, and were of analytical purity or reagent grade.

# CHAPTER 1

**allylic oxidation reactions using palladium as a catalyst**

## 3.  Results and discussion

To carry out the allylic oxidation reactions on nitrogen compounds, you first need to follow all the synthetic steps for producing 2-hydroxy-naphthoquinones, which are described below.

### 3.1. Synthesis of nor-lapachol using Hooker's oxidation method

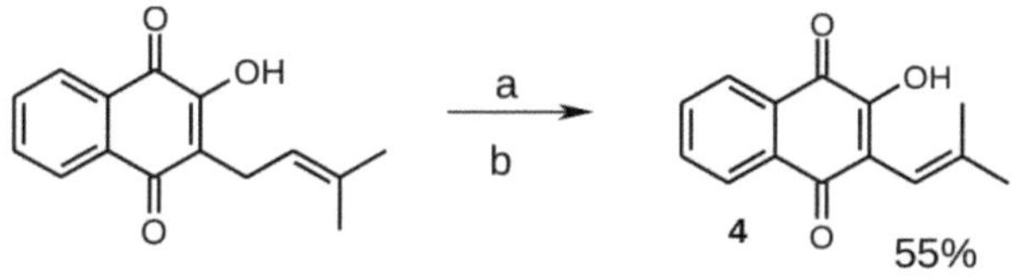

lapachol, 6

(a) $Na_2CO_3$, $H_2O_2$ 30%; (b) NaOH, $CuSO_4$; △

Scheme 15: Synthesis of nor-lapachol from lapachol

The nor-lapachol synthesized by the Hooker oxidation methodology was obtained with a yield of 55%, and the product was characterized by comparison in CCDA using the pure nor-lapachol standard.

A silica gel chromatographic column was used to purify nor-lapachol, using hexane and ethyl acetate (8:2) as the eluent. Orange crystals with a melting point between 120-122 °C were obtained. The infrared spectrum (p. 68) showed an absorption characteristic of axial deformation of the OH bond at 3364 cm$^{-1}$ , absorptions at 1661 cm$^{-1}$ and 1644 cm$^{-1}$ which are characteristic of axial deformation of the C=O bond of conjugated ketones. There was also an absorption at 1626 cm$^{-1}$ , characteristic of the C=C bond of alkene, which was attributed to the double bond of the nor-lapachol side chain, as well as absorptions at 2969 cm$^{-1}$ and 2927 cm$^{-1}$ , which were attributed to axial deformations of the C-H bonds of alkene and alkanes in this chain, respectively. Also in the infrared spectrum, an intense absorption was observed at 1276 cm$^{-1}$ , attributed to an axial deformation of the C-O bond. Pairs of absorptions characteristic of axial deformations of C=C bonds in aromatics were observed at 1592 cm$^{-1}$ and 1453 cm$^{-1}$ , in addition to a characteristic absorption at 730 cm$^{-1}$ which was attributed to an out-of-plane deformation for the C-H bond of the ortho-substituted benzene ring.

The following is a mechanistic proposal for the formation of nor-lapachol by the Hooker oxidation process. This proposal is based on Lee's work (Lee, 1995), which shows that nor-lapachol is formed by the loss of carbon from position 2, as shown in Scheme 16:

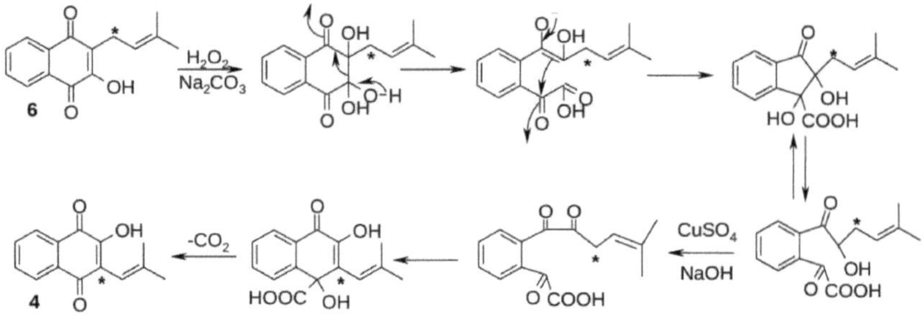

Diagram 16: Mechanistic proposal (Lee, 1995) for the formation of nor-lapachol

This proposal is based on the whole process, without any modification of the reaction conditions, and is therefore applicable to lapachol.

## 3.2 Synthesis of nor-lapachol by condensation in hydrochloric acid and acetic acid

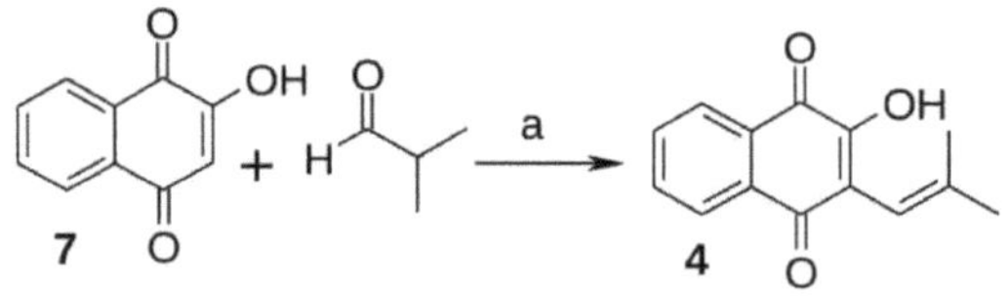

a: HOAc; $HCl_{(aq)}$; refluxo; 45 %

Scheme 17: Synthesis of nor-lapachol by acid condensation

This reaction is carried out at a temperature of 70-75 °C, the lausone is dissolved in acetic acid and then the aldehyde and hydrochloric acid are mixed simultaneously (Hooker, 1936). The reaction system is then darkened and left to react for 1.5 hours. After this time, the entire contents of the flask were added to a Becker with crushed ice and neutralized with a 1% NaOH solution. The final product was obtained with a yield of 45 %.

## 3.3 Synthesis of nor-lapachol using the modified Kopanski method (Kopanski, 1987).

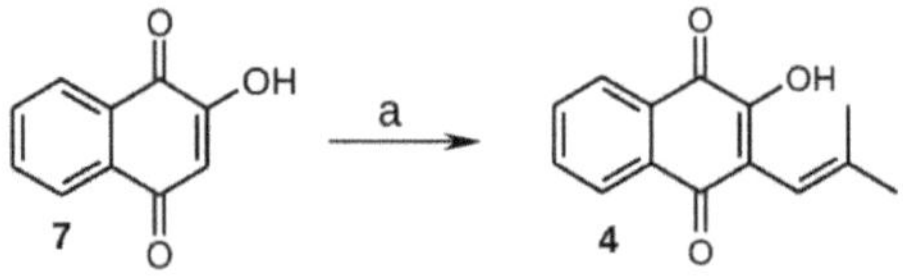

(a) PhMe; beta alanine; HOAc; isobutyraldehyde; reflux 1h; 96 %

Diagram 18: Synthesis of nor-lapachol using β-alanine

Using a Dean-Stark apparatus, the water formed in the reaction is removed and the equilibrium is shifted towards the production of alkene. The reaction finishes within an hour and purification is carried out by acid-base extraction, where a solution of sodium carbonate is prepared, the toluene is not evaporated, the entire contents of the flask are added to a separating funnel and extraction is carried out with the sodium carbonate solution. During extraction, a kind of floating liquid is formed, and the basic solution should be added until the water is as clear as possible. All the purple liquid is collected and neutralized with diluted phosphoric acid. Wait 24 hours, decant and vacuum filter. The product obtained is already pure enough and there is no need for column chromatography.

### 3.4. Vinyl ester preparation reaction

As mentioned earlier, the formation of the vinyl ester (Scheme 8, p. 12) was not very successful, with a yield of 16 %. However, the fact that the solvent used in the reaction was n-hexane could have been the reason for the inefficiency of the process, as the DDQ was not completely solubilized. So it was decided to change the reaction solvent, and what was obtained is shown in Scheme 19:

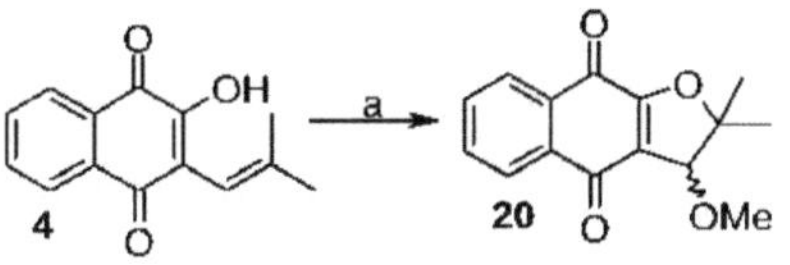

a: MeOH; DDQ; refluxo          45 %

Diagram 19: Formation of the five-membered vinyl ester

Observation of the NMR spectrum[1] H, (300 MHz, CDCl3, p. 69), showed the appearance of a methoxyl group as a simplet at δ 3.52 ppm referring to three hydrogens. There were also two simpletons belonging to the two terminal methyls. The fact that they appear with different chemical shifts is because they are linked to a prochiral carbon, so they have different chemical environments. The spectrum of[13] C (p. 72) shows the appearance of 15 types of carbon, showing that methanol has been added to the structure of the product.

The infrared spectrum carried out on a KBr tablet (p. 73) shows the methyl bands at 2983 and 2934 cm$^{-1}$ and the methoxyl at 1364 cm$^{-1}$ . The melting point of this ester was 92-93°C.

Methanol behaved like a nucleophile if added to the structure of the molecule, and a mechanistic proposal was made, as can be seen in Diagram 20:

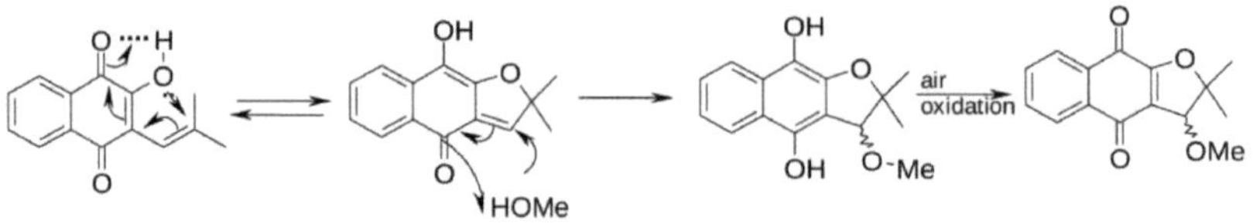

Diagram 20: Mechanistic proposal for the formation of **20**

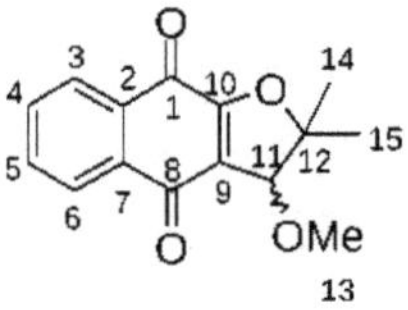

Table 1: Spectroscopic data of **20**

|  | $^1$H | $^{13}$C |
|---|---|---|
| 1 | - | 181,08 |
| 2 | - | 127,53 |
| 3 | 7,57(d, 1H, $J_{3,4}$=2,4 Hz) | 127,00 |
| 4 | 7,62(dt, 1H, $J_{4,3}$=2,4; $J_{4,5}$=1,5 e $J_{4,6}$=1,2 Hz) | 129,34 |
| 5 | 7,67(dt, 1H, $J_{5,4}$=1,5 e $J_{5,6}$=7,5 Hz) | 132,44 |
| 6 | 8,07(dt, 1H, $J_{6,5}$=7,5 e $J_{6,4}$=1,2 Hz) | 134,52 |
| 7 | - | 131,11 |
| 8 | - | 175,92 |
| 9 | - | 116,41 |
| 10 | - | 170,87 |
| 11 | 4,46 (s,1H) | 95,71 |
| 12 | - | 59,15 |
| 13 | 3,52 (s, 3H) | 83,76 |
| 14 | 1,45 (s, 3H) | 26,68 |
| 15 | 1,60 (s, 3H) | 20,49 |

Table 2 shows the spectroscopic data of the six-membered vinyl ester obtained in the reaction in which hexane was the solvent.

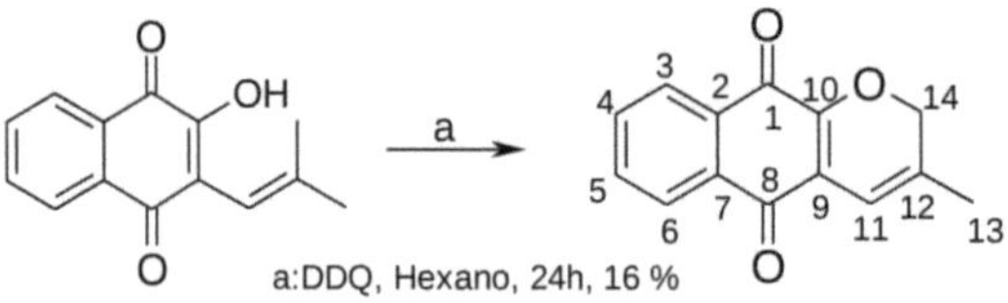

Table 2: Spectroscopic data of **11**

| | $^{1}H$ |
|---|---|
| 1 | - |
| 2 | - |
| 3 | 8,06 (t, 1H, $J_{3,4}$=1,4 Hz) |
| 4 | 7,66 (dt, 1H, $J_{4,3}$=1,4 e $J_{4,5}$=2,2 Hz) |
| 5 | 7,66 (dt, 1H, $J_{5,4}$=2,2 e $J_{5,6}$=3,2 Hz) |
| 6 | 8,04 (dd, 1H, $J_{6,5}$=3,2 Hz) |
| 7 | - |
| 8 | - |
| 9 | - |
| 10 | - |
| 11 | 6,43 (dd, 1H, $J_{11,13}$=1,2 e $J_{11,14}$=1,2 Hz) |
| 12 | - |
| 13 | 1,83 (dd, 3H, $J_{13,11}$=1,2 e $J_{13,14}$=1,4 Hz) |
| 14 | 4,90 (dd, 2H, $J_{14,11}$=1,6 e $J_{14,13}$=1,2 Hz) |

The$^{1}$ H spectrum of (**11**) can be seen on page 100, showing a coupling between hydrogen 13 and 11, which is a homoallylic coupling, with a $J$ value of 1.2 Hz. There is also a coupling of the same type between hydrogen 11 and 14 with a constant value of 1.6 Hz.

### 3.5. **Preparation of naphthoquinones with amine functionality**

Cyclo-oxidations were initially investigated with hydroxylated derivatives rather than nitrogenated ones, because they are more abundant in our laboratory, since the reaction to synthesize these nitrogenated compounds goes through two successive steps and the final yield is reduced. The first stage consists of substituting the methyl group hydrogen using dimethyl sulphate, as shown in Scheme 21.

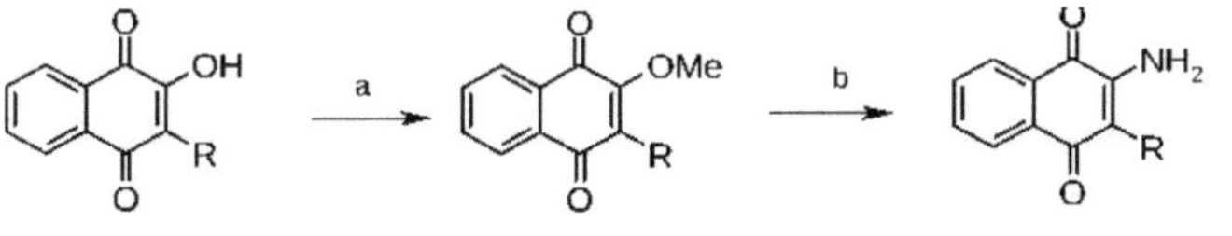

OH    a    OMe    b    $NH_2$

4.R=CHC($CH_3$)$_2$
5.R=(CH)$_2$CH$_2$CH$_3$

4'.R=CHC($CH_3$)$_2$
5'.R=(CH)$_2$CH$_2$CH$_3$

4a.R=CHC($CH_3$)$_2$ 58 %
5a.R=(CH)$_2$CH$_2$CH$_3$ 31 %

a = $K_2$ $CO_3$ , $Me_2$ $SO_4$ , propanone, stir, 8 h b = $NH_4$ OH, MeOH, stir, 24 h

Diagram 21: Preparation of nitrogen compounds

Step **a is** carried out without isolating the products (**4'**) and (**5'**), a suggested mechanism is shown below, Diagram 22.

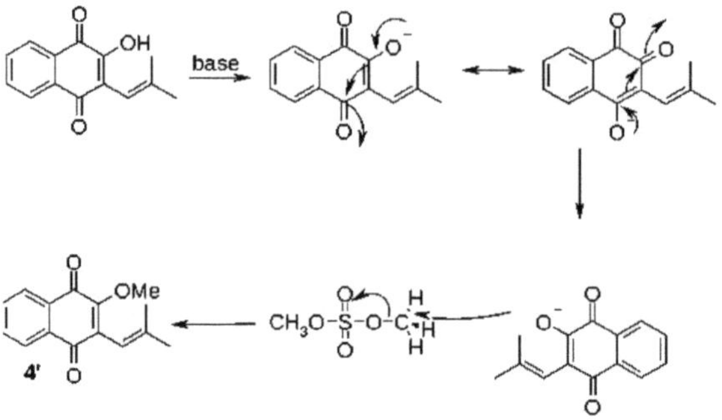

Diagram 22: Mechanistic proposal for methoxylation with (CH3)2SO4

When the methoxyl group was replaced by the amine group of the molecule (**4'**), excess dimethyl sulphate was used, which led to the formation of a monomethylated product in the amine group (**4b**), with a yield of 80% (Figure 9). This product was isolated and characterized using IR and elemental analysis techniques.

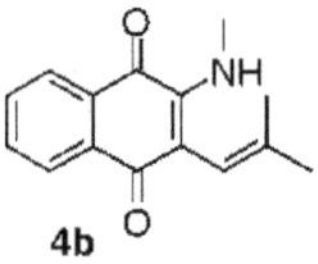

Figure 8: Monomethylation product

The product (**4b**) was subjected to X-ray diffraction analysis Figure 11, by the Crystallography and Molecular Modeling Laboratory, in cooperation with Prof[a] . Valéria Rodrigues dos Santos Malta and her master's student Jademilson Celestino dos Santos, both from the Federal University of Alagoas (UFAL).

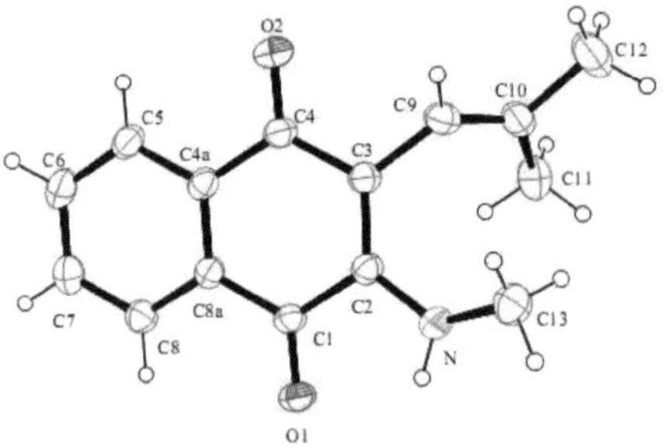

Figure 9: ORTEP diagram of N-methyl-amino-norlapachol

Using the SciFinder Scholar® program, it was found that this molecule is new to the literature. Its melting point was 111-112 °C.[3]

### 3.6. **Tosylation of nor-lapachol**

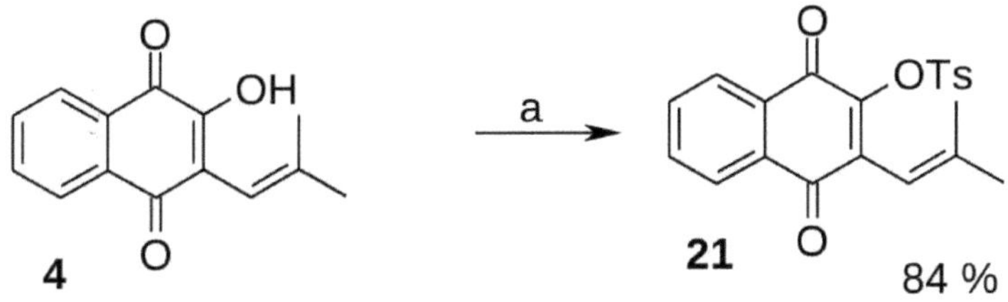

a: propanone; $Na_2CO_3$ ; TsCl; ultrasound;

Scheme 23: Nor-lapachol tosylation reaction

The tosylation of nor-lapachol was carried out in order to facilitate the entry of the amine group in position 2, since tosyl is a better leaving group compared to methoxyl, and the reaction for the production of amino-norlapachol takes place more easily. The tosylation process was carried out in an ultrasound machine for a period of 1 hour and 10 minutes. The final product is crystalline with a melting point between 100-102 °C, and the yield of this reaction is 84 %. The NMR analysis of $^{1}$ H (CDCl3, 200 MHz, p. 77) and $^{13}$ C (CDCl3, 50 MHz, p. 78) respectively, shows that the tosylation took place successfully, as the simplet of the methyl linked to the aromatic was found at δ 2.48 ppm, as well as the displacements of the aromatic rings and the integration showing a total of eight aromatic hydrogens.

Table 3: NMR data $^{1}$ H and $^{13}$ C of **21** (CDCl3, 200 MHz)

3 The *N* -methyl-amino-nor-lapachol molecule has been subjected to X-ray diffraction analysis and the article is under submission to the Journal: J.Mol.Struct.

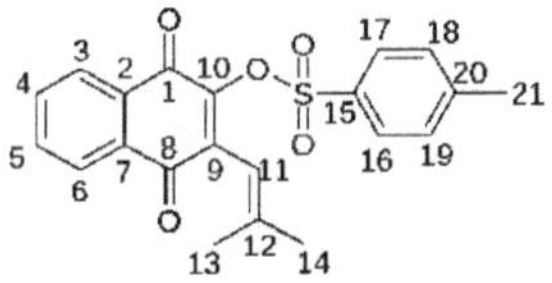

| | $^1$H | $^{13}$C |
|---|---|---|
| 1 | - | 184,71 |
| 2 | - | 137,31 |
| 3 | 8,15 (dt, 1H, $J_{3,4}$=10,0; $J_{3,5}$=7,5 e $J_{3,6}$=2,8) | 134,17 |
| 4 | 7,77 (dt, 1H, $J_{4,3}$=10,0 e $J_{4,5\,e\,4,6}$=8,5 Hz) | 128,19 |
| 5 | 7,71 (dt, 1H, $J_{5,3}$=7,5; $J_{5,4}$=8,5 e $J_{5,6}$=10,0 Hz) | 129,63 |
| 6 | 8,09 (dt, 1H, $J_{6,5}$=10,0; $J_{6,4}$=8,5 e $J_{6,3}$=2,8 Hz) | 134,10 |
| 7 | - | 134,43 |
| 8 | - | 178,91 |
| 9 | - | 126,82 |
| 10 | - | 146,68 |
| 11 | 5,78 (s, 1H) | 113,50 |
| 12 | - | 137,10 |
| 13 | 1,73 (s, 3H) | 26,60 |
| 14 | 1,56 (s, 3H | 21,88 |
| 15 | | 145,43 |
| 16 | 7,86 (d, 1H, $J_{16,19}$=8,0 Hz) | 129,73 |
| 17 | 7,86 (d, 1H, $J_{17,18}$=8,0 Hz) | 130,80 |
| 18 | 7,37 (d, 1H, $J_{18,17}$=8,1 Hz) | 126,75 |
| 19 | 7,37 (d, 1H, $J_{19,16}$=8,1 Hz) | 126,68 |
| 20 | | 131,86 |
| 21 | 2,48 (s, 3H) | 21,71 |

## 3.7. Formation reaction of α-xyloidone

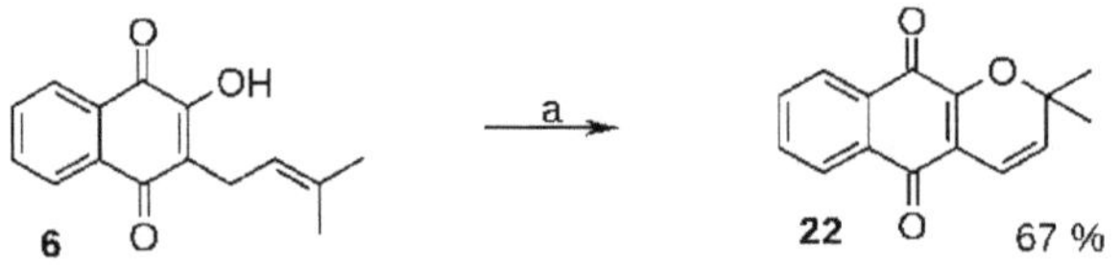

a: Pd(OAc)$_2$ ; Cu(OAc)$_2$ ; MeOH;

Scheme 24: Obtaining α-xyloidone

This methodology was developed by Heck (Heck, 1985) with the aim of acetoxylating alcohols in allylic positions. However, despite the fact that there is no hydroxyl in an allylic position, this idea was used to check the behavior of naphthoquinone against reagents containing palladium as a catalyst in oxidations. Using lapachol (**6**) as the starting product, the condition shown above was applied, and at the end of the reaction the solvent was evaporated under low pressure in a rotaevaporator and then vacuum filtered with a thin layer of diatomaceous earth. Separation was carried out with ethyl acetate (4x30 ml). The organic phases were combined and concentrated under reduced pressure in a rotaevaporator. Purification was carried out by chromatographic column using silica gel as the fixed phase and an elution system of hexane and ethyl acetate (8:2). Yellowish crystals were obtained with a mass of 806 mg, i.e. a yield of 67%. Methodologies for obtaining α-xyloidone are well known in the literature, with yields of 65 % (Tapia, 2001), 55 % (Lee, 2004), 77 % (Kumar, 2008), and between 10-53 % depending on the solvent and reaction time (Ribeiro, 2008).The melting point was obtained and recorded as 139-140 °C. The product (**22**) was characterized by IR (p. 79), NMR of[1] H (CDCl3, 200 MHz, p. 80) and[13] C (CDCl3, 50 MHz, p. 81), and it was found that there was no acetoxylation as initially proposed, because the spectrum of[1] H did not show the simplet of the acetate group, the spectrum of[13] C showed the appearance of 14 types of carbons, which does not match if there was acetylation. The spectrum of[1] H showed a singlet with a ratio of six hydrogens, which is equivalent to two methyls with the same chemical environment. Two duplets were observed at δ 5.70 ppm with *J=10* Hz and the other at δ 6.62 ppm and *J=10* Hz, corresponding to two CH double bonds. The IR spectrum did not show the hydroxyl band, which suggests that cyclization may have taken place. It was possible to observe the CH bands at 2975 cm[-1] corresponding to the two methyls.

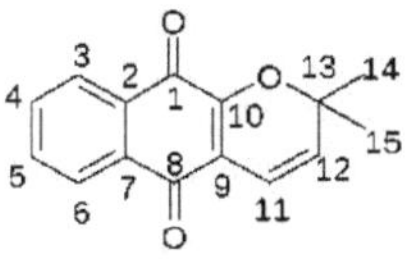

Table 4: NMR data of **22** (CDCl3, 200 MHz)

| | $^1$H | $^{13}$C |
|---|---|---|
| 1 | - | 183,21 |
| 2 | - | 132,28 |
| 3 | 8,06 (dt, 1H, $J_{3,4}$=3,5 e $J_{3,5}$=2,5 Hz) | 132,90 |
| 4 | 7,64 (dd, 1H, $J_{4,3}$=3,5; $J_{4,6}$=2,7 e $J_{4,6}$=2,0 Hz) | 116,83 |
| 5 | 7,68 (dt, 1H, $J_{5,3}$=2,5; $J_{5,4}$=2,7 e $J_{5,6}$= 4,0 Hz) | 119,22 |
| 6 | 8,03 (dd, 1H, $J_{6,5}$=4,0 e $J_{6,4}$=2,0 Hz) | 132,28 |
| 7 | - | 132,83 |
| 8 | - | 181,21 |
| 9 | - | 134,59 |
| 10 | - | 153,79 |
| 11 | 5,70 (s, 1H, $J_{11,12}$=10 Hz) | 127,59 |
| 12 | 6,67 (d, 1H, $J_{12,11}$=10 Hz) | 135,36 |
| 13 | - | 81,82 |
| 14 | 1,53 (s, 3H) | 29,76 |
| 15 | 1,53 (s, 3H) | 29,76 |

An explanation for measuring the non-acetoxylation of the lapachol molecule is shown in Scheme 25. The reaction system consists of palladium acetate and copper II acetate and the reaction was carried out in an open environment. A catalytic cycle is formed between the palladium acetate which is added to the lapachol side chain along with the addition of an acetate group to the neighboring carbon (Tsuji, 1995). Cyclization occurs by expelling the acetate group and acetic acid is formed *in situ*, and soon after the palladium is abandoned by elimination, which now enters the catalytic cycle as Pd(0) and is re-oxidized by copper acetate to Pd(OAc)2 Scheme 25.

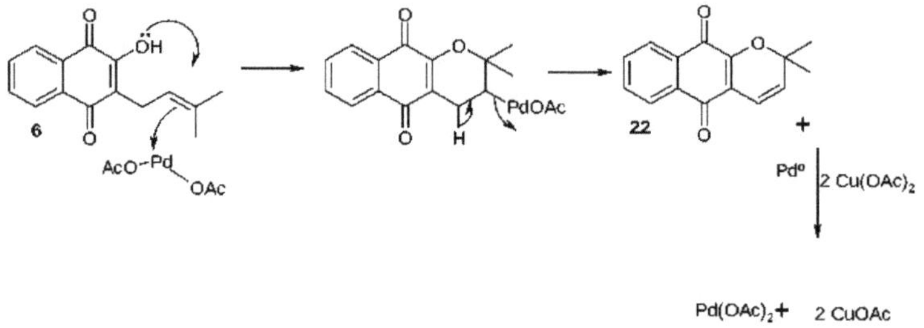

Diagram 25: Catalytic cycle of palladium proposed for the formation of α-xyloidone **22**

### 1.8.  Acetoxylation of nor-lapachol by Heck's methodology (Heck, 1985)

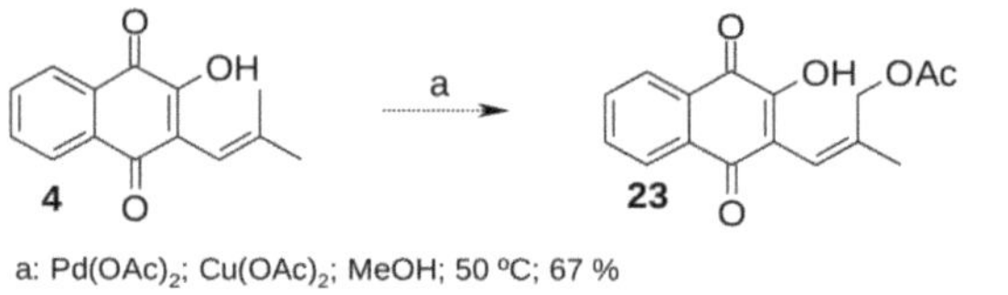

a: Pd(OAc)$_2$; Cu(OAc)$_2$; MeOH; 50 °C; 67 %

Scheme 26: acetoxylation of nor-lapachol

This same type of reaction was carried out for nor-lapachol (**4**) in order to also evaluate the behavior of this non-nitrogenous quinone in relation to the reagents used. Following the same experimental procedure, an orange-colored product was isolated in 67% yield after using a chromatographic column with silica gel as the fixed phase and the elution system being hexane and ethyl acetate (9:1).

### 1.9.  Acetoxylation of nor-isolapachol by Hansson's methodology (Hansson, 1990)

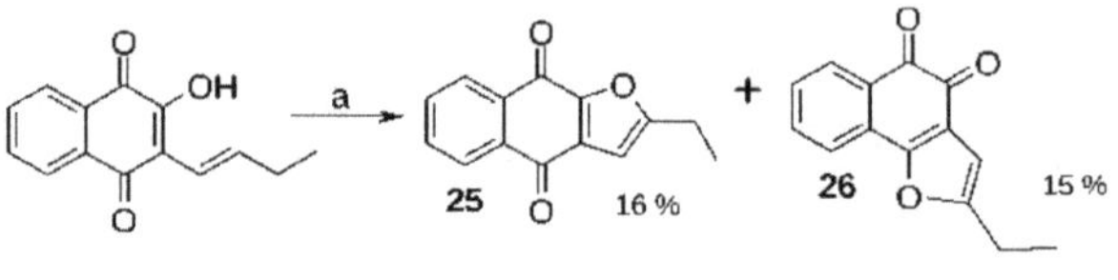

a: 0.05 mol % Pd(OAc)$_2$ ; MnO$_2$ 200 mol %; HOAc

Diagram 27: Cyclization of the iso-nor-lapachol side chain

Following the methodology in diagram 27, two products were obtained after purification on a chromatographic column: a dark red crystal, characteristic of beta cyclization in naphthoquinones, and a yellow crystal, characteristic of alpha cyclization. The melting point of the red crystal (**26**) was measured at 140-141 °C and the yellow crystal (**25**) at 132-133 °C. The IR data shown in the spectrum (p. 82) shows that compound (**25**) has carbonyl absorption at 1672 cm 1 and no broad hydroxyl band, which suggests that the molecule has been cyclized. The absorption of the carbonyl of compound (**26**) (p. 87) was at 1699 cm$^{-1}$ and also shows no hydroxyl band. A possible mechanism for what might have happened is shown in Diagram 28.

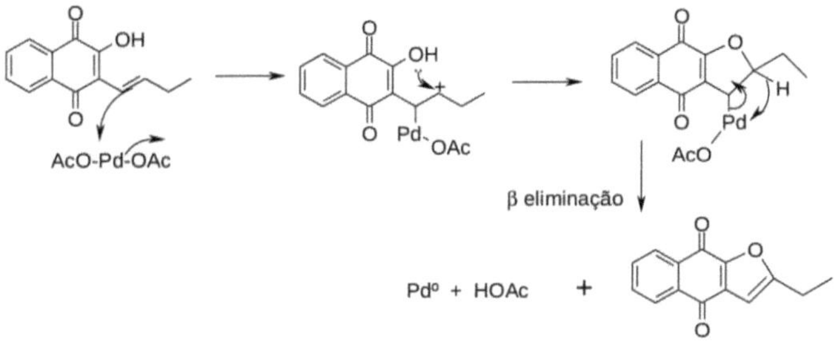

scheme 28: proposed mechanism of α cyclization.

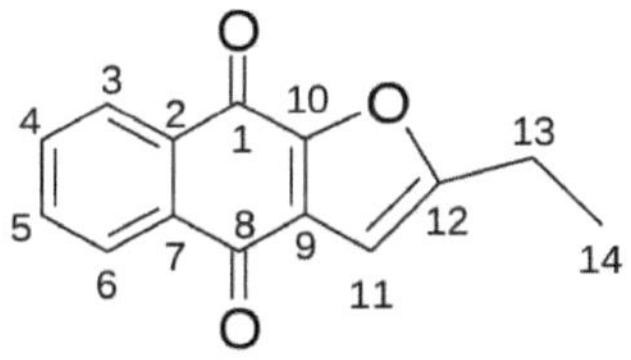

Table 5 shows the spectroscopic data obtained for the α (**25**) isomer

|  | $^1$H | $^{13}$C |
|---|---|---|
| 1 | - | - |
| 2 | - | 126,83 |
| 3 | 7,39(ddd, 1H, $J_{3,4}$=1,8; $J_{3,5}$=1,5 e $J_{3,6}$=0,9 Hz) | 126,77 |
| 4 | 7,35(dt, 1H, $J_{4,3}$=1,8; $J_{4,5}$=2,1 e $J_{4,6}$=1,2 Hz) | 133,05 |
| 5 | 7,59 (dt, 1H, $J_{5,3}$=1,5; $J_{5,4}$=2,1 e $J_{5,6}$=2,4 Hz) | 133,53 |
| 6 | 7,71 (ddd, 1H, $J_{6,3}$=0,9; $J_{6,4}$=1,2 e $J_{6,5}$=2,4 Hz) | 133,79 |
| 7 | - | 132,52 |
| 8 | - | 165,89 |
| 9 | - | 131,75 |
| 10 | - | 173,19 |
| 11 | 6,61 (t, 1H, $J_{11,13}$=0,9 Hz) | 103,50 |
| 12 | - | - |
| 13 | 2,84 (dq, 2H, $J_{13,11}$=0,9 e $J_{13,14}$=7,5 Hz) | 21,77 |
| 14 | 1,35 (t, 3H, $J_{14,13}$=7,5 Hz) | 11,57 |

Similarly, the formation of its isomer can also be proposed, as shown in Diagram 29:

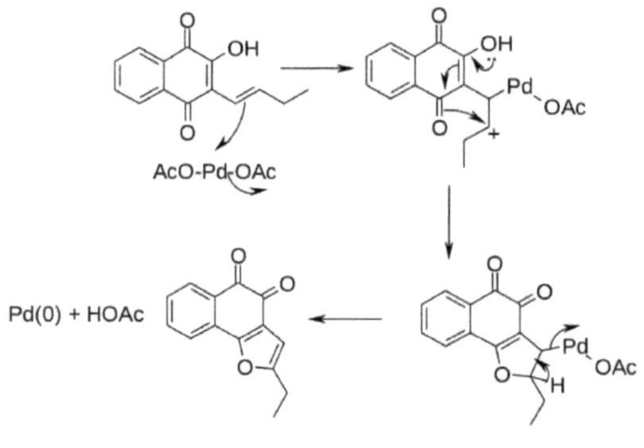

Diagram 29: Proposed mechanism of β-isomer formation

The spectroscopic data for compound (**26**) are shown in Table 6.

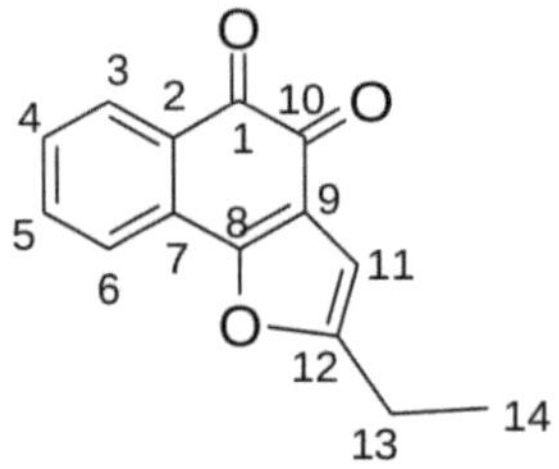

Table 6: Spectroscopic data of **26**

| | $^1$H | $^{13}$C |
|---|---|---|
| 1 | - | 180,62 |
| 2 | - | 122,35 |
| 3 | 7,39(dd, 1H, $J_{3,4}$=2,7 e $J_{3,5}$=0,9 Hz) | 121,91 |
| 4 | 7,35(ddd, 1H, $J_{4,3}$=2,7; $J_{4,5}$=1,2 e $J_{4,6}$=0,6 Hz) | 129,70 |
| 5 | 7,59(ddd, 1H, $J_{5,3}$=0,9; $J_{5,4}$=1,2 e $J_{5,6}$=6,3 Hz) | 130,31 |
| 6 | 7,98(dd, 1H, $J_{6,4}$=0,6 e $J_{6,5}$=6,3 Hz) | 135,27 |
| 7 | - | 128,60 |
| 8 | - | 161,40 |
| 9 | - | 128,44 |
| 10 | - | 174,31 |
| 11 | 6,39(t, 1H, $J_{11,13}$=0,9 Hz) | 102,90 |
| 12 | - | 159,46 |
| 13 | 2,71(dq, 2H, $J_{13,11}$=0,9 e $J_{13,14}$=8,4 Hz) | 21,20 |
| 14 | 1,28(t, 3H, $J_{14,13}$=8,4 Hz) | 11,56 |

## 1.10. Hansson's acetoxyaction for amino-nor-lapachol (Hansson, 1990)

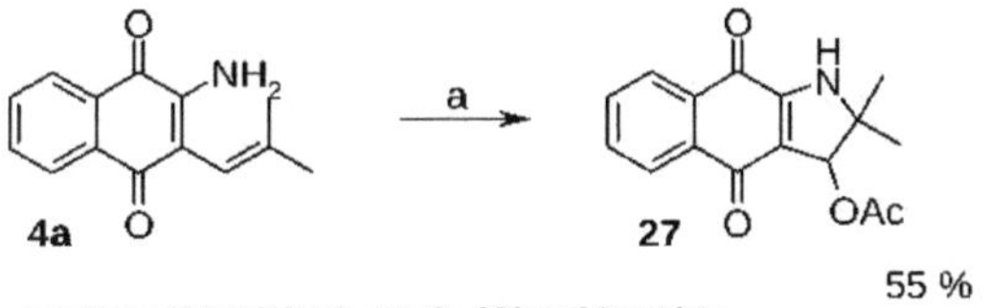

Diagram 30: Formation reaction of **27**

Now using a nitrogenous naphthoquinonic compound, amino-nor-lapachol (**4a**), an acetoxylation in the allylic position was expected. After purification by column chromatography, a yellow-colored compound (**27**) with a melting point of 188189 °C was isolated. The spectrum of $^{13}$C (CDCl3, 125 MHz, p. 94) shows the appearance of 16 types of carbon atom, which when compared using the NMR spectrum $^1$H (CDCl3, 500 MHz, p. 93) shows the appearance of an acetate group, as a singlet at δ 2.11 ppm, there are also two singlets at δ 1.53 and 1.60 ppm, referring to the terminal methyl groups of the molecule's side chain. A simplet with hydrogen integration at δ 6.07 ppm is characteristic of CH linked to the acetate group, and a signal with low intensity at δ 5.29 ppm is compatible with N-H. Diagram 31 suggests the formation of molecule (**27**).

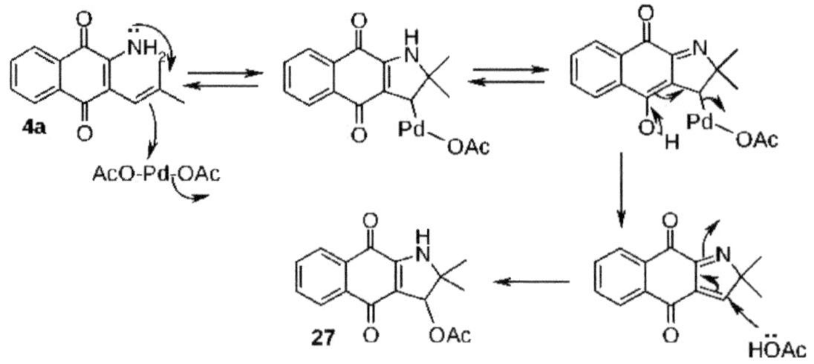

Diagram 31: Proposed mechanism for the formation of **27**

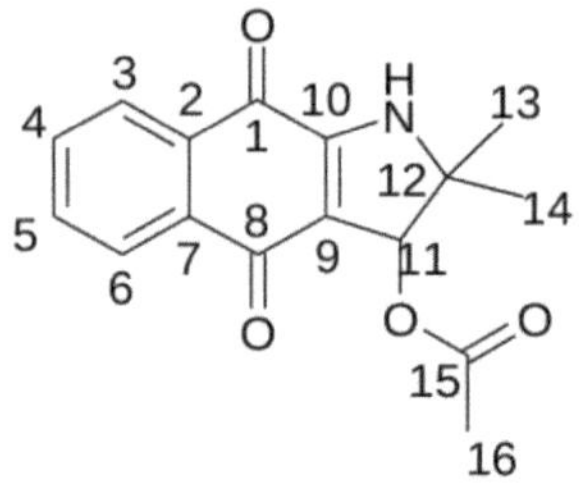

Table 7: NMR data of **27** in (CDCl3)

|  | $^1$H (500 MHz) | $^{13}$C (125 MHz) |
|---|---|---|
| 1 | - | 180,44 |
| 2 | - | 112,74 |
| 3 | 8,13 (dd, 1H, $J_{3,4}$=5,5 e $J_{3,5}$=1,0 Hz) | 134,43 |
| 4 | 7,64 (dd 1H, $J_{4,3}$=5,5; $J_{4,5}$=5,5 e $J_{4,6}$=1,5 Hz) | 132,64 |
| 5 | 7,68 (dd, 1H, $J_{5,4}$=5,5 e $J_{5,6}$=5,5 Hz) | 129,46 |
| 6 | 7,71 (dd, 1H, $J_{6,4}$=1,5 e $J_{6,5}$=5,5 Hz) | 124,97 |
| 7 | - | 126,97 |
| 8 | - | 174,64 |
| 9 | - | 131,26 |
| 10 | - | 170,44 |
| 11 | 6,07 (s, 1H) | 76,54 |
| 12 | - | 95,14 |
| 13 | 1,60 (s, 3H) | 20,54 |
| 14 | 1,53 (s, 3H) | 20,51 |
| 15 | - | 170,70 |
| 16 | 2,11 (s, 3H) | 25,90 |

Table 8 shows the elemental analysis data for products 22, 20, 23, 4b and 20:

Table 8: Elemental analysis (theoretical/found)

| Composto | % (C)t/e | % (H)t/e | % (O)t/e |
| --- | --- | --- | --- |
| 22 | 74,99/72,95 | 5,03/4,82 | 19,98/22,22 |
| 20 | 69,76/68,39 | 5,46/5,46 | 24,78/26,13 |
| 25 | 74,33/77,40 | 4,46/5,36 | 21,22/17,23 |
| 26 | 74,33/77,82 | 4,46/5,58 | 21,22/16,58 |
| 4b | 74,67/74,93 | 6,27/6,14 | 13,26/12,94 |
| 20 | 69,76/69,16 | 5,46/5,61 | 24,78/25,22 |

Analysis carried out at the Department of Fundamental Chemistry -DQF- of the Federal University of Pernambuco

As can be seen, only compounds 4b and 20 were characterized using this technique.

# 4. EXPERIMENTAL

## 4.1. Synthesis of nor-lapachol using Hooker's oxidation method (Hooker, 1936)

In a 1 L round-bottomed flask, 12.1 g of lapachol (50 mmol), 5 g of potassium carbonate (36.2 mmol) and 200 mL of a 1:1 solution of distilled water and ethanol were added. The reaction mixture was heated to approximately 50° C and then 10 mL of a 30% hydrogen peroxide solution was slowly added. The mixture was kept stirring and after 12 hours it became discolored. The ethanol was evaporated in a rotaevaporator at reduced pressure, the flask containing the mixture was placed in an ice bath and then small portions of sodium metabisulphite were added, followed by 100 mL of a 25 % sodium hydroxide solution and 250 mL of a solution containing 50 g of copper sulphate (CuSO4) alternately. The mixture was heated for 15 minutes. After standing for an hour, the mixture was filtered through celite and the purple-colored filtrate was acidified with a 10 % H3PO4 solution until it became yellowish in color, followed by extraction with ethyl acetate (5 x 50 mL). The organic phases were combined into a single phase, which was dried with anhydrous sodium sulphate and concentrated under reduced pressure. The product was subjected to column chromatography using hexane 8:2 acetate as eluent, giving an orange-colored product, which was compared to the pure nor-lapachol standard, with a yield of ~55%.

## 4.2. Synthesis of nor-lapachol using the modified Kopanski methodology (Kopanski, 1987)

Starting from 10 mmol of lausone (**7**) and 20 mmol of isobutyraldehyde, using 0.25 g of β-alanine and 0.15 g of glacial acetic acid. In a round-bottomed flask add a few small pieces of porcelain and place the lausone, 100 ml of toluene, 0.25 g of β-alanine, mix the aldehyde with the acetic acid separately and then add this mixture to the flask, then place it on a blanket with pieces of porcelain and leave it to reflux in toluene for an hour in an open system. With this collection system called Dean-Stark, the water formed is retained in the well of the flask and the equilibrium is shifted towards the production of alkene. The reaction finishes within an hour and purification is carried out by acid-base extraction, where a sodium carbonate solution must be prepared, the toluene is not evaporated, and the entire contents of the flask are placed in a separatory funnel and extracted there with the sodium carbonate solution. During the extraction, a kind of floating liquid is formed and the basic solution should be added until the color of the water is as clear as possible. All the purple liquid is collected and neutralized with diluted phosphoric acid. Wait 24 hours, decant and vacuum filter. The product obtained is completely pure and there is no need for column chromatography. The yield obtained was 96% and the product was compared in

CCDA with a pure standard of norlapachol.

### 4.3. Synthesis of nor-lapachol by condensation with hydrochloric acid

In a 100 ml round-bottomed flask, 2 g of lausone and 35 ml of acetic acid were added. The system was left at a temperature of approximately 80 °C, and then 2 ml of HCl and 5 ml of isobutyraldehyde were added at the same time. The system went dark immediately after the addition of these two substances, and after 1.5 hours the extraction process was carried out. The contents of the flask were added to a 500 ml beaker filled with crushed ice. A 1:1 benzene-ether extraction was carried out and the organic phase was separated using a separating funnel. The organic phase was treated with a 1% NaOH solution to transform norlapachol into a sodium salt, and then the aqueous phase was treated with dilute phosphoric acid and left to stand for 24 hours. The product was compared by analytical thin layer chromatography (ADLC) with the pure nor-lapachol standard, obtaining a yield of 44%.

### 4.4. Preparation of the five-membered vinyl ester

Starting with 3 mmol of nor-lapachol (684 mg) and 3 mmol of DDQ (681 mg), a reflux was carried out using methanol as a solvent, and the reaction ended thirty minutes later. At the end of the reaction, the methanol was evaporated and then dichloromethane was added in order to vacuum filter it and retain the (insoluble) DDQ. Purification was carried out by chromatographic column using hexane and ethyl acetate 8:2. The product obtained was subjected to NMR analysis[1] H and[13] C and its melting point was recorded at 92-93 °C, its yield was 45 %. The product has an orange color and a crystalline structure.

### 4.5. Tosylation of nor-lapachol

1 mmol of *nor-lapachol* and 1.5 molar equivalents of tosyl chloride were used in 35 mL of acetone and 0.5 mmol of sodium carbonate. The system was left under ultrasound until all the *nor-lapachol was* consumed, monitored by ccda, and it was found that the reaction finished in approximately 1 hour and 10 minutes. A vacuum filtration was carried out and methanol was used to rinse the funnel, all the organic phase was collected and rotaevaporated under reduced pressure and the formation of yellowish crystals could already be seen, with a melting point of between 100-102 °C and a yield of 84 %.

**IV(KBr)**v max 3065, 2983, 2934, 2831, 1688, 1649, 1620, 1594, 1460, 1396, 1364, 1095, 964, 724 cm[-1.]

**NMR[1] H (CDCl3, 300 MHz, ppm)** 7.57(d, 1H, *J3.4*=2.4 Hz); 7.62(dt, 1H, *J4.3*=2.4; *J4.5*=1.5 and *J4.6*=1.2 Hz); 7.67(dt, 1H, *J5.4*=1.5 and *J5.6*=7.5 Hz); 8.07(dt, 1H, *J6.5*=7.5

and *J6.4=1.2 Hz*); 4.46 (s,3H); 3.52 (s, 3H); 1.45 (s, 3H); 1.60 (s, 3H).

**NMR$^{13}$ C (CDCl3, 75 MHz, ppm)** 175.92; 127.53; 127.00; 129.34; 132.44; 134.52; 131.11; 181.08; 116.41; 170.87; 95.71; 59.15; 83.76; 26.68; 20.49.

### 4.6. Synthesis of amino nor-lapachol

The product 4$^a$ was synthesized in situ and the intermediate 4' was not isolated due to its instability. For this purpose, 1 mmol of nor-lapachol, 1.2 equivalents of dimethyl sulphate and 1.5 equivalents of potassium carbonate were used in approximately 30 ml of acetone. The system is left to stir constantly for two hours, after which the solvent is evaporated and 30 ml of methanol and 10 ml of ammonium hydroxide are added. This process takes approximately 3 hours and the product can be obtained by adding water to the reaction medium and extracting with (3 x 15 ml) of dichloromethane. A chromatographic column is made using hexane/ethyl acetate 8:2.

### 4.7. Synthesis of α-xyloidone 22

We used 1 mmol of lapachol (**6**) in 2.5 mL of methanol, 1 mL of distilled water, 0.02 mmol of palladium acetate and 2.5 mmol of cupric acetate pentahydrate. The reaction is kept stirring with the system open for a period of 24 hours. The isolation is carried out by column chromatography with silica gel (fixed phase) and the elution system (mobile phase) is hexane/ethyl acetate (8:2). Yellow crystals with a melting point of 139-140 °C were obtained. The yield of this reaction was 67%.

**IV (KBr)** v max 2975, 1674, 1646, 1593, 1570, 1449, 1414, 1368, 1329, 1274, 1210, 1189, 1159, 1135, 968, 717 cm$^{-1}$ .

**NMR$^1$ H (CDCl3, 200 MHz, ppm)** 8.06 (dt, 1H, *J3.4=3.5 and J3.5=2.5 Hz*); 7.64 (dd, 1H, *J4.3=3.5; J4.5=2.7 and J4.6=2.0 Hz*); 7.68 (dt, 1H, *J5.4=2.7 and J5.6= 4.0 Hz*); 8.03 (dd, 1H, *J6.5=4.0 and J6.4=2.0 Hz*); 5.70 (s, 1H, *J11.12=10 Hz*); 6.67 (d, 1H, *J12.11=10 Hz*); *1*.53 (s, 3H); 1.53 (s, 3H).

**NMR$^{13}$ C (CDCl3, 50 MHz, ppm)** 181.21; 132.28; 132.90; 116.83; 119.22; 132.28; 132.83; 183.21; 134.59; 153.79; 127.59; 135.36; 81.82; 29.76; 29.76.

### 4.8. Hansson acetoxylation for nor-isolapachol

Using the same experimental procedure, 3 mmol of iso-nor-lapachol, 684 mg, and 0.05 mol % of Pd(OAc)2 (3.36 mg), 200 mol % of MnO2 (522 mg) and 30 mL of HOAc were used. The MnO2 was activated by heating at approximately 150 °C for 30 minutes, then the acid and MnO2 were left to stir for 30 min. at a gentle heat of 50 °C. At the end of the reaction one hour later, vacuum filtration is carried out with a thin layer of preparative plate silica or diatomaceous earth. After purification by chromatographic column, two compounds were obtained in crystalline form, one colored dark red and the other yellow. The melting point of the red crystal (**26**) was measured at 132-133 °C and the yellow crystal (**25**) had its f.p. in the range 140-141 °C. Their yields were 16 % and 15 % respectively, as shown below.

**IV (KBr)** v max 3120, 2986, 2934, 1672, 1592, 1533, 1458, 1427, 1373, 1334, 1276, 1218, 1195, 1160, 959, 715 cm$^{-1}$ .

**NMR[1] H (CDCI3, 300 MHz, ppm)** 7.39(ddd, 1H, *J3*.4=1.8 Hz); 7.35(dt, 1H, *J4*.3=1.8; *J4*.5=2.1 and *J4*.6=1.2 Hz); 7.59 (dt, 1H, *J5*.4=2.1 and *J5*.6=2.4 Hz); 7.71 (ddd, 1H, *J6*.4=1.2 and *J6*.5=2.4 Hz); 6.61 (t, 1H, *J11*.13=0.9 Hz); 2.84 (dq, 2H, *J13*.11=0.9 and *J13*.14=7.5 Hz); 1.35 (t, 3H, *J14*.13=7.5 Hz).

**NMR[13] C (CDCI3, 75 MHz, ppm)** 126.83; 126.77; 133.05; 133.53; 133.79; 132.52; 165.89; 131.75; 173.19; 103.50; 21.77; 11.57

. **IR (KBr)**v max 2980, 2937, 2878, 1699, 1671, 1577, 1522, 1485, 1422, 1379, 1342, 1311, 1273, 1214, 1115, 895, 769, 467 cm 1 .

**NMR[1] H (CDCI3, 300 MHz, ppm)** 7.39(dd, 1H, *J3*.4=2.7 Hz); 7.35(dt, 1H, *J4*.3=2.7; *J4*.5=0.9 and *J4*.6=0.6 Hz); 7.59(dd, 1H, *J5*.4=0.9 and *J5*.6=6.3 Hz); 7.98(dd, 1H, *J6*.4=0.6 and *J6*.5=6.3 Hz); 6.39(t, 1H, *J11*.13=0.9 Hz); 2.71(dq, 2H, *J13*.11=0.9 and *J13*.14=8.4 Hz); 1.28(t, 3H, *J14*.13=8.4 Hz).

**NMR[13] C (CDCI3, 75 MHz, ppm)** 180.62; 122.35; 121.91; 129.70; 130.31; 135.27; 128.60; 161.40; 128.44; 174.31; 102.90; 159.46; 21.20; 11.56.

### 4.9. Hansson's acetoxylation of amino-nor-lapachol

Following the same experimental procedure for this type of acetoxylation, we used 2 mmol of amino-nor-lapachol (454 mg) from (**4a**). We were able to isolate a product that was characterized as (**24**) as yellow crystals with a melting point in the range of 188-189 °C and its yield was 55 %.

**IV (KBr)** v max 3245, 2962, 2932, 2970, 1663, 1642, 1591, 1458, 1330, 1259, 1208, 1048, 955, 891, 720 cm$^{-1}$ .

**NMR[1] H (CDCI3, 500 MHz, ppm)** 8.13 (dd, 1H, J3.4=5.5 and J3.5=1.0 Hz); 7.64 (dd 1H, J4.3=5.5; J4.5=5.5 and J4.6=1.5 Hz); 7.68 (dd, 1H, J5.4=5.5 and J5.6=5.5 Hz); 7.71 (dd, 1H, J6.4=1.5 and J6.5=5.5 Hz); 6.07 (s, 1H); 1.60 (s, 3H); 1.53 (s, 3H); 2.11 (s, 3H)

**NMR**[13] **C (CDCl3, 175 MHz, ppm)** 170.44; 129.46; 95.14; 131.26; 112.74; 126.97; 124.97; 131.26; 180.44; 132.64; 76.54; 134.43; 20.54; 20.51; 174.64; 25.90.

# CHAPTER 2

**Synthesis of DiMers**

## 5.  Synthesis of dimers

The methodology of (Kopanski, 1987) shown in Figure 10, uses β-alanine and acetic acid as catalysts and benzene as a solvent, obtaining the product (**28**) with a yield of 94 %.

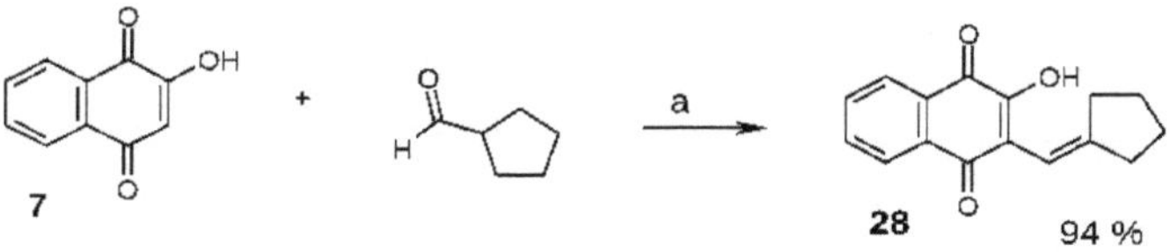

a: benzene; beta-alanine; HOAc; $N_2$ ;

Figure 10: Kopanski condensation reaction

It was then decided to apply this methodology, as shown in Figure 11, to the synthesis of norlapachol (**4**).

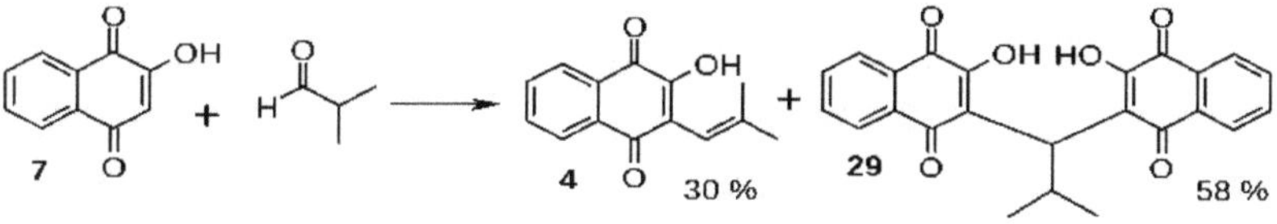

Figure 11: Condensation of lausone with isobutyraldehyde

When this synthesis was carried out (Figure 11), it was found by CCDA that nor-lapachol was not the only product being formed (accompanied by a nor-lapachol pattern), and a very polar orange-colored product was observed. The isolation and characterization of this product was carried out and confirmed by NMR analysis of[1] H, which showed a double duplet at 1.55 ppm referring to the isopropyl group, different from a singlet, which should appear if the alkene had been formed, by the symmetry of the molecule, it was verified that the carbon displacements referred to 14 types of carbons, However, the essential information to provide certainty that the alkene was not formed was the non-appearance of the methylenic CH signal, which would appear at approximately 5.90 ppm, and therefore the suggestion that best fitted the spectral data was that it was a dimer, as shown in Figure 11.

The melting point was measured and it was found that it decomposed between temperatures of 237-238. The yield of this dimer was 58%, while that of nor-lapachol was

30%. It was decided to modify the methodology, where acetic acid was the catalyst, and it was discovered that the use of this acid in excess favored the almost exclusive formation of the dimer, and the yield was now 89 %. A literature search revealed that molecules similar to this one were being used in tests against HIV (Mahindra, 2001), (Mahindra, 2002). It was therefore decided to develop a homologous series of these dimers using various available aldehydes (Table 9), as well as to carry out the spectroscopic characterizations and report these previously unpublished data.

29  R= $CH(CH_3)_2$
29a R= $(CH_2)_2CH_3$
29b R= $(CH_2)_4CH_3$
29c R= $CHC_6H_5$
29d R= $C_6H_4OCH_3$
29e R= H
29f R= $C_6H_4NO_2$
29g R= $C_6H_4OH$
29h R= $C_6H_4F$
29i R= $CH(CH_2)_2(CH_3)_2$
29j R= $CH_3$

Table 9: Year-on-year series ratio

| CÓDIGO | PONTO DE FUSÃO | PESO MOLECULAR | FÓRMULA MOLECULAR |
|---|---|---|---|
| 29 | 185-186 (dec) | 402 | $C_{24}H_{18}O_6$ |
| 29[a] | 206-207 (dec) | 402 | $C_{24}H_{18}O_6$ |
| 29b | 256 (dec) | 430 | $C_{26}H_{22}O_6$ |
| 29c | 273 (dec) | 436 | $C_{27}H_{16}O_6$ |
| 29d | 224-225 | 466 | $C_{28}H_{18}O_7$ |
| 29e | 251 (dec) | 360 | $C_{21}H_{12}O_6$ |
| 29f | 257 (dec) | 481 | $C_{27}H_{15}O_8N$ |
| 29g | 134-135 | 452 | $C_{27}H_{16}O_7$ |
| 29h | 136-137 | 454 | $C_{27}H_{15}O_6F$ |
| 29i | 183 (dec) | 430 | $C_{26}H_{22}O_6$ |
| 29j | 175-176 | 374 | $C_{22}H_{14}O_6$ |

Using the SciFinder Scholar® program, it was found that molecules **29d**, **29f**, **29g** and **29h** are new to the literature. All these molecules were used, in partnership with Dr.[a] . Tânia Maria Sarmento da Silva, to carry out biological tests, one of which was against human topoisomerase II.

A probable reaction mechanism is shown in Diagram 32.

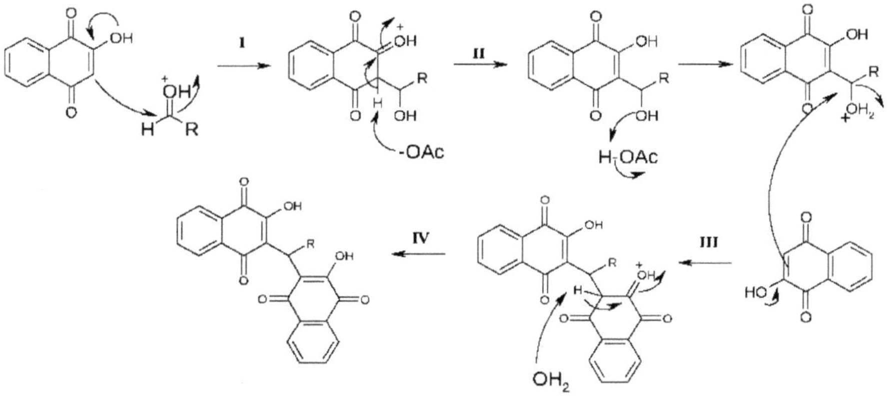

Diagram 32: Proposed dimerization mechanism

The removal of β-alanine resulted in a delay of 5 days to consume all the starting product. However, the yield of the final product was not significantly altered.

It was decided to change the solvent and observe its influence on the reaction, using dichloromethane, and the yield was better than using benzene, but no better than using excess AcOH (Table 10).

Table 10: Conditions and yields of the dimer formation reactions

| Entry | Aldeido | Condition 1 (%) | Condition 2 (%) | Condition 3 (%) | Condition 4 (%) |
|---|---|---|---|---|---|
| 1 | n-butyraldehyde | 24 h (36) | 23 h (80) | X | 2 h (55) |
| 2 | isobutyraldehyde | 24 h (58) | 24 h (89) | X | X |
| 3 | paraformaldehyde | X | 6 h (75) | X | X |
| 4 | acetaldehyde | X | 1 h 15 min. (72) | X | X |
| 5 | 2-ethyl-butyraldehyde | X | 2 h 15 min. (68) | X | X |
| 6 | Hexanaldehyde | X | 35 min. (69) | X | X |
| 7 | Benzaldehyde | X | 4 h (93) | X | X |
| 8 | anisic aldehyde | X | 50 min. (94) | 5 days (96) | X |

| 9 | 4-fluor-benzaldehyde | X | 1 h 15 min (98). | 5 days (98) | X |
| 10 | 4-nitro-benzaldehyde | X | 1 h 15 min. (98) | X | X |
| 11 | 4-hydroxy benzaldehyde | X | 55 min. (83) | X | X |

Condition 1: benzene as solvent; Condition 2: acetic acid as solvent; Condition 3: reaction without the addition of β-alanine; Condition 4: dichloromethane as solvent.

All these methodologies were proposed in order to observe the reaction profile and find out which is the best system to favor the formation of alkenes. It can be seen from Table 10 that the aldehydes which have a lower electronic density in the carbonyls are the ones which gave the best yields, since the initial nucleophilic attack of lausone (**7**) on the carbonyl is favored. As can be seen from the detailed study of Table 10, which relates the types of yields with the conditions used, even without carrying out all the experiments, some useful conclusions can be drawn, such as, when using benzene *versus* acetic acid, the yields of dimers with condition 2 are better, because the extremely acidic medium activates the aldehyde carbonyls. When comparing conditions 1 and 2 with 4, it can be seen that time was the factor that most favored condition 4, but its yield compared to 2 suggests that the use of HOAc is still much better. As for the condition comparing the use or not of β-alanine, it can be seen that the reaction time was very different, justifying the use of the catalyst.

Observing that water is eliminated, step **11** to **11 I**, Scheme 32, we can then use a well-known principle, which is Le Chatelier's principle. This is where water is removed from the reaction medium so that the equilibrium is shifted towards the formation of more product. So we used an apparatus that was able to retain the water.

**5.1 Synthesis of 29: 2,2'-(2-methylpropylidene)bis[3-hydroxy-1,4-naphthalenedione]**

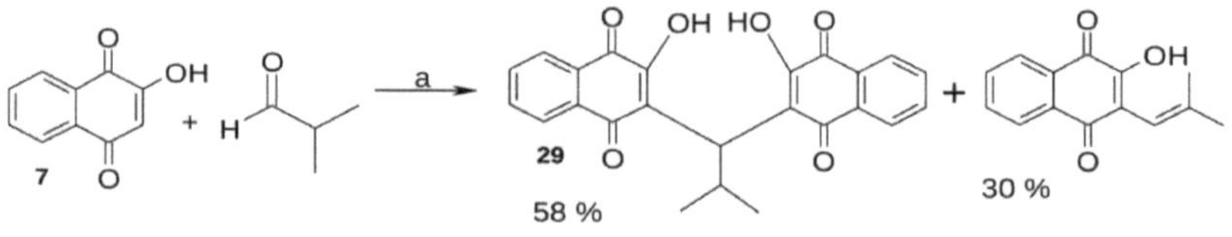

**Diagram 33: Condensation with isobutyraldehyde**

The dimer (**29**) was obtained using the condition in which benzene was the reaction solvent. The reaction was carried out in a nitrogen atmosphere and kept at 50 °C for 24 hours. The isolation was carried out by removing the benzene under a rotaevaporator at low pressure, after which a chromatographic column was made using silica gel as the fixed phase and hexane/ethyl acetate (9:1). The orange-colored isolated product was subjected to NMR analysis at[1] H and[13] C and its decomposition point was measured at 185-186 °C. Its yield was 58 %. The infrared spectrum analysis (p. 95) it was possible to observe an absorption band in the region of compounds with hydroxyls, the absorption was broad and strong at 3452 cm$^{-1}$ , characteristic of an asymmetric stretching of the -OH bond, a very characteristic absorption of the -C-H sp bond[3] referring to the alkyl groups at 2962 cm$^{-1}$ the carbonyl absorption, as expected, appeared as two strong bands referring to the carbonyls at 1670 cm$^{-1}$ and 1627 cm$^{-1}$ . Analysis of the NMR spectrum of[1] H (*DMSOd6*, 300 MHz, p. 96) in the region from 0.6 to 2.0 ppm, shown on (p. 124), reveals a duplex. 124), there is a duplet at δ 0.78 ppm (*J=6Hz*) referring to the methyls of the isopropyl group, which is coupled with the CH group showing (septet, δ 2.95 ppm, 1H, *J*= 6.3 Hz), and which is coupled with a CH δ 4.80 ppm (duplet ,1H, *J=11* Hz). The APT spectrum[13] C (*DMSOd6*, 75 MHz, p. 99) shows 14 types of carbon, which means that there are overlapping signals, as the molecule is symmetrical.

Table 11: NMR data of **29** in *DMSOd6* (δ =ppm, 300 MHz)

| | $^1$H | $^{13}$C |
|---|---|---|
| 1 | - | 183,14 |
| 2 | - | 133,14 |
| 3 | 7,90 (d, 2H, $J_{3,4}$=7 Hz) | 131,82 |
| 4 | 7,62 (dt, 2H, $J_{4,3}$=7,0 e $J_{4,5}$=5,5 Hz) | 125,69 |
| 5 | 7,71 (dt, 2H, $J_{5,4}$=7,2 e $J_{5,6}$=6,6 Hz) | 124,89 |
| 6 | 7,83 (d, 2H, $J_{6,5}$=7,5 Hz) | 130,79 |
| 7 | - | 133,71 |
| 8 | - | 183,09 |
| 9 | - | 123,48 |
| 10 | - | 163,44 |
| 11 | 4,93 (d, 1H, $J_{11,12}$=10,8 Hz) | 36,39 |
| 12 | 2,98 (m, 1H, $J_{12,13}$=6,3 e $J_{12,11}$=10,8 Hz) | 24,16 |
| 13 | 0,78 (d, 6H, $J_{13,12}$=6,3 Hz) | 21,65 |

Observation of table 11 shows that the carbon displacements correspond to 13 different types of carbon, but the molecule as a whole has 24 carbons. Due to the symmetry of the molecule, there is an overlap of the signals and what can be seen is that the displacements of 11 carbons have been suppressed. The displacements of the aliphatic carbons 11, 12 and 13 appear in high field, showing that they are not alkenes, which are also formed in this reaction.

**5.2 Synthesis of 29a: 2,2'-(2-butylidene)bis[3-hydroxy-1,4-naphthalenedione]**

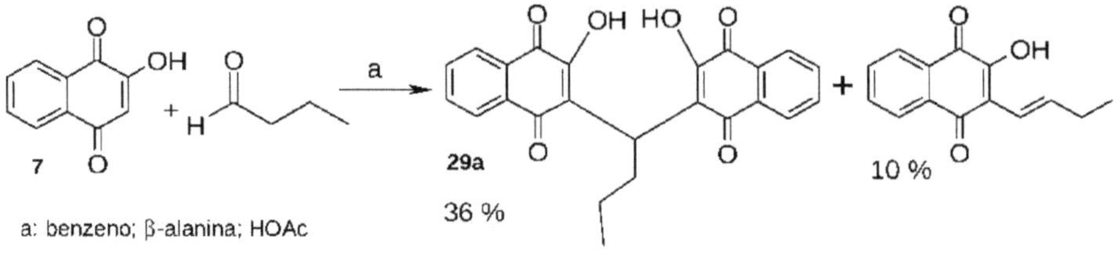

Esquema 34: Condensação com n-butiraldeído

For the synthesis of dimer **29a**, the methodology in which benzene is the reaction solvent was used. Isolation is carried out by removing the benzene under a rotaevaporator at low pressure, followed by column chromatography using silica gel and hexane/ethyl acetate (9:1) to remove the part of the corresponding alkene that is formed. The polarity was increased to isolate an orange-colored product with a decomposition point of 206-207 °C, with a yield of 36 %. The IR spectrum (p. 100) shows a broad, strong hydroxyl band at 3452 cm$^{-1}$ two absorption bands corresponding to the asymmetric deformation of the terminal methyl of the aliphatic part at 2958 and 2927 cm 1. The absorption of the carbonyl group which resonates with the hydroxyl group appears at 1627 cm$^{1}$ while the other carbonyl at 1666 cm$^{-1}$ . The$^{1}$ H spectrum (p. 102) shows the coupling relationship between the aliphatic hydrogens and the$^{13}$ C spectrum shows the appearance of 14 types of carbon. Table 12 shows the spectral data obtained.

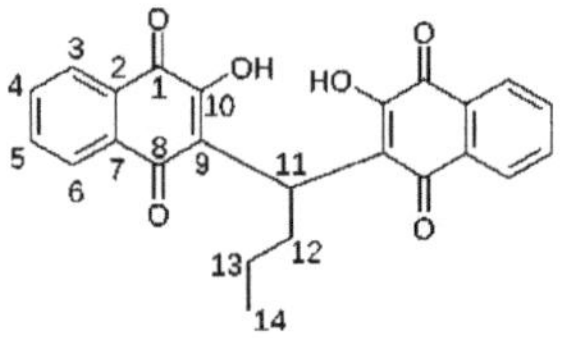

Table 12: **29a** NMR data in *DMSOd6* (δ =ppm, 300 MHz)

|  | $^1$H | $^{13}$C |
|---|---|---|
| 1 | - | 184,12 |
| 2 | - | 133,24 |
| 3 | 7,91 (d, 2H, $J_{3,4}$=7,0 Hz) | 131,92 |
| 4 | 7,62 (dt, 2H, $J_{4,3}$=7,0 e $J_{4,5}$=7,5 Hz) | 125,05 |
| 5 | 7,72 (dt, 2H, $J_{5,4}$=7,5 e $J_{5,6}$=7,5 Hz) | 125,74 |
| 6 | 7,83 (t, 2H, $J_{6,5}$=7,5 Hz) | 130,72 |
| 7 | - | 133,90 |
| 8 | - | 182,77 |
| 9 | - | 123,54 |
| 10 | - | 163,82 |
| 11 | 5,31 (t, 1H, $J_{11,12}$= 8,1 Hz) | 38,66 |
| 12 | 1,54 (dd, 2H, $J_{12,11}$=8,1 e $J_{12,13}$=7,5 Hz) | 30,87 |
| 13 | 1,15 (dt, 2H, $J_{13,12}$=7,5 e $J_{13,14}$=7,2 Hz) | 21,09 |
| 14 | 0,83 (t, 2H, $J_{14,13}$=7,2 Hz) | 14,11 |

## 5.3 Synthesis of 29b: 2,2'-(hexylidene)bis[3-hydroxy-1,4-naphthalenedione]

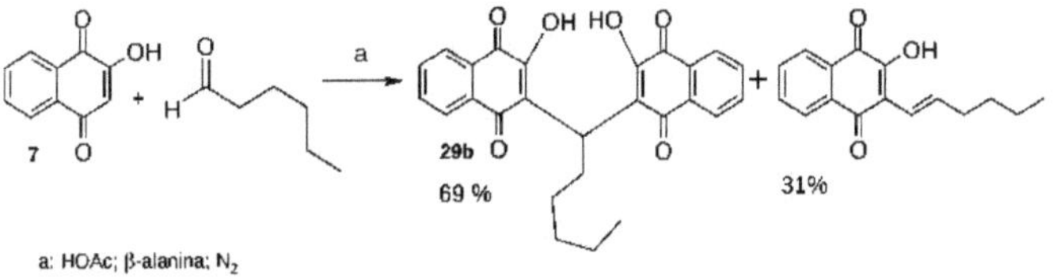

Scheme 35: Condensation with n-hexanaldehyde

The synthesis of (**29b**) was carried out using acetic acid as the solvent.

Table 10 shows that the use of acetic acid favors the formation of dimers. The system was kept at a temperature of 50 °C in a nitrogen atmosphere for 35 minutes. Isolation was carried out by adding distilled water to the flask and filtering the precipitate, which was washed thoroughly with distilled water to remove excess HOAc and hexanaldehyde. The final product was an orange-colored solid, obtained in a yield of 69% and with a decomposition point of 256 °C. Analysis of the[1] H spectrum carried out in (DMSO *d6*, 300

45

MHz, p. 104) shows a triplet at 0.9 ppm referring to the terminal methyl of the aliphatic chain with the neighboring CH2, a triplet with a shift of 5.2 ppm referring to the CH that joins the two molecules. IR absorptions (p. 108) show two bands at 2927 and 2854 cm$^{-1}$ which are due to the asymmetric stretching of CH sp$^3$ . The same absorptions of the carbonyls observed for (**29a**) are observed in (**29b**), which are the bands at 1666 and 1627 cm$^{-1}$ . Analysis of the spectrum of$^{13}$ C, (DMSO *d6*, p. 107) shows the appearance of 16 types of carbon in the structure, of which six are saturated aliphatic carbons.

Table 13: NMR data of **29b** in *DMSOd6* (δ =ppm, 300 MHz)

| | $^1$H | $^{13}$C |
|---|---|---|
| 1 | - | 183,64 |
| 2 | - | 133,10 |
| 3 | 7,92 (dd, 2H, $J_{3,4}$=7,5 e $J_{3,5}$=0,9 Hz) | 131,98 |
| 4 | 7,63 (dt, 2H, $J_{4,3}$=7,5; $J_{4,5}$= 7,2 e $J_{4,6}$=1,2 Hz) | 125,04 |
| 5 | 7,72 (dt, 2H, $J_{5,4}$=7,2; $J_{5,6}$=7,5 e $J_{5,3}$=0,9 Hz) | 125,73 |
| 6 | 7,85 (dd, 2H, $J_{6,5}$=7,5 e $J_{6,4}$=1,2 Hz) | 130,56 |
| 7 | - | 133,90 |
| 8 | - | 182,87 |
| 9 | - | 123,68 |
| 10 | - | 162,70 |
| 11 | 5,18 (t, 1H, $J_{11,12}$=8 Hz) | 31,37 |
| 12 | 1,99 (m, 2H) | 29,10 |
| 13 | 1,18 (m, 2H) | 28,76 |
| 14 | 1,18 (m, 2H) | 27,61 |
| 15 | 1,18 (m, 2H) | 22,08 |
| 16 | 0,77 (t, 3H, $J_{16,15}$=6,9 Hz) | 13,97 |

## 5.4 Synthesis of 29c: 2,2'-(phenylmethylene)bis[3-hydroxy-1,4-naphthalenedione]



46

Diagram 36: Condensation with benzaldehyde

This reaction was carried out using HOAc as the solvent and β-alanine as the catalyst. Isolation was carried out by adding distilled water to the reaction flask, after which the precipitate was filtered. The precipitate is washed thoroughly with distilled water to remove excess HOAc and benzaldehyde. The final product is a yellow-colored solid, obtained in 93% yield and with a decomposition point of 273 °C. The spectrum of[1] H (300 MHz, DMSOd6, p. 109) shows a H-methyl simplet at δ 6.70 ppm (s, 1H). The phenyl signals appeared as a multiplet at 7.14 ppm (m, 5H). The IR spectrum shown on (p. 111) shows the appearance of the hydroxyl band at 3320 cm$^{-1}$ , a sharp and strong band at 1570 cm$^{-1}$ referring to the stretching of the C=C bond of benzene. The spectrum of[13] C (DMSOd6, 75 MHz, p. 112) shows the appearance of only one aliphatic carbon peak with a shift of δ 33.08 ppm, and the other peaks appeared with the C sp shift[2] , as well as two peaks referring to carbonyls at δ 182.63 and 183.84 ppm.

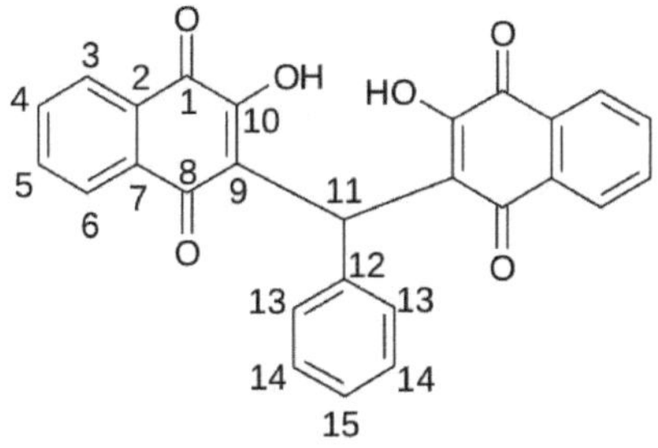

|  | $^1$H | $^{13}$C |
|---|---|---|
| 1 | - | 183,84 |
| 2 | - | 133,27 |
| 3 | 7,72 (d, 2H, $J_{3,4}$=7,5 Hz) | 124,90 |
| 4 | 7,76 (t, 2H, $J_{4,3\,e\,4,5}$=7,5 Hz) | 132,05 |
| 5 | 7,67 (t, 2H, $J_{5,4\,e\,5,6}$= 7,5 Hz) | 125,84 |
| 6 | 7,89 (d, 2H, $J_{6,5}$=7,5 Hz) | 130,89 |
| 7 | - | 133,98 |
| 8 | - | 182,63 |
| 9 | - | 122,84 |
| 10 | - | 164,27 |
| 11 | 6,70 (s, 1H) | 33,08 |
| 12 | - | 141,27 |
| 13 | * | 125,22 |
| 14 | * | 127,83 |
| 15 | * | 126,85 |

* 7,15 (m, 5H)

Table 14: **29c** NMR data in *DMSOd6* (δ =ppm, 300 MHz)

Synthesis of 29d: 2,2'-(para-methoxyphenylmethylene)bis[3-hydroxy-1,4-naphthalenedione]

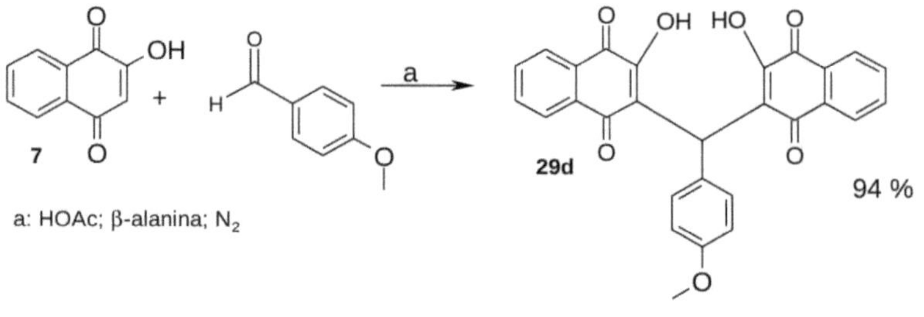

Scheme 37: Condensation with anisic aldehyde

Isolation is carried out by adding distilled water to the flask and filtering the precipitate, which is washed abundantly with distilled water to remove excess HOAc and anisic aldehyde. The final product is a yellow-colored solid, obtained in 94% yield and with a melting point of 224-225 °C. Analysis of the NMR spectrum[1] H (DMSO *d6*, 300 MHz, p. 113) shows the displacement of the methoxyl group as a singlet at δ 3.69 ppm, and an integration system showing that it may indeed be the dimer (**29d**). The displacement of the methine CH is observed as a simplet at δ 5.94 ppm, (s,1H). The methoxyl can be evidenced by the simplet at δ 3.69 ppm and by the spectrum of[13] C (*DMSOd6*, 75 MHz, p. 115) as a peak shifted δ 54.94 ppm. The IR spectrum (p. 116) shows absorptions of 3394 cm$^{-1}$ which is the band of the hydroxyl groups, two strong and sharp absorptions belonging to the two different carbonyls, 1666 cm$^{-1}$ referring to the carbonyl conjugated only with the aromatic and 1639 cm$^{-1}$ referring to the carbonyl conjugated with the hydroxyl group.

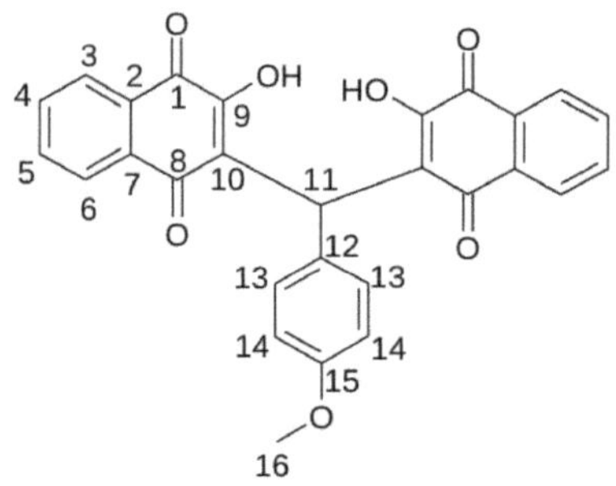

Table 15: **29d** NMR data in *DMSOd6* (δ =ppm, 300 MHz)

| | $^1$H | $^{13}$C |
|---|---|---|
| 1 | - | 183,74 |
| 2 | - | 132,70 |
| 3 | 7,98 (dd, 2H, $J_{3,4}$=7,0 e $J_{3,5}$=1,2 Hz) | 125,64 |
| 4 | 7,77 (dt, 2H, $J_{4,3}$=7,0; $J_{4,6}$=1,2 e $J_{4,5}$=7,5 Hz) | 132,21 |
| 5 | 7,77 (dt, $J_{5,3}$=1,2; $J_{5,4}$=7,5 e $J_{5,6}$=7,2 Hz) | 126,11 |
| 6 | 7,98 (dd, $J_{6,5}$=7,2 e J6,4=1,2 Hz) | 129,85 |
| 7 | - | 133,16 |
| 8 | - | 181,28 |
| 9 | - | 155,90 |
| 10 | - | 123,63 |
| 11 | 5,94 (s, 1H) | 37,31 |
| 12 | - | 134,73 |
| 13 | 6,78 (dd, 2H, $J_{13,14}$=8,7 Hz) | 129,37 |
| 14 | 7,18 (dd, 2H, $J_{14,13}$=8,7 Hz) | 113,10 |
| 15 | - | 157,34 |
| 16 | 3,69 (s, 3H) | 54,94 |

Synthesis of 29e: 2,2'-(methylene)bis[3-hydroxy-1,4-naphthalenedione]

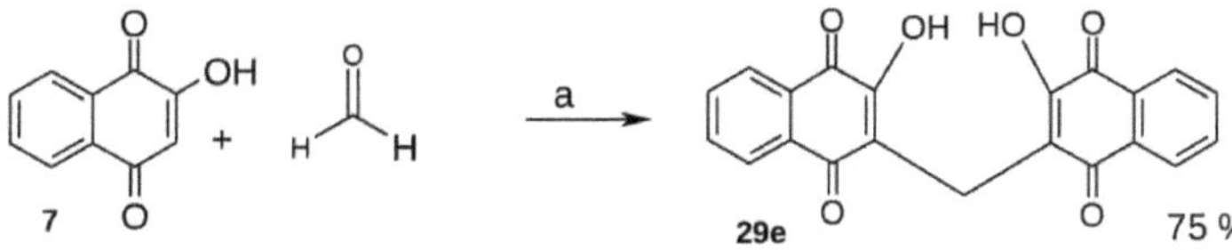

a: HOAc; β-alanina; $N_2$

Diagram 38: Condensation with formaldehyde

Because formaldehyde contains a large amount of water, paraformaldehyde was used to carry out this reaction. Isolation was carried out by adding distilled water to the flask and filtering the precipitate, which was washed abundantly with distilled water to remove the excess HOAc and paraformaldehyde. The final product was a yellow-colored solid, obtained in a yield of 75 % and with a degradation point of 251 °C, which was compared with that in the literature, which is 249-251 dec. The spectrum of[1] H (DMSO *d6*, 300 MHz, p. 117) shows the displacement signal of methyl CH2 as a singlet at δ 3.75 ppm and the displacements of the aromatic hydrogens in the region of δ 7.72 to 7.96 ppm. The spectrum of[13] C (*DMSOd6*, 75 MHz, p. 118) shows only eleven types of carbon and the displacement of this same CH2 δ 17.94 ppm. The IR spectrum (p. 119) shows the absorption of CH2 as a band at 3070 cm$^{-1}$ , the hydroxyl band at 3452 cm$^{-1}$ .

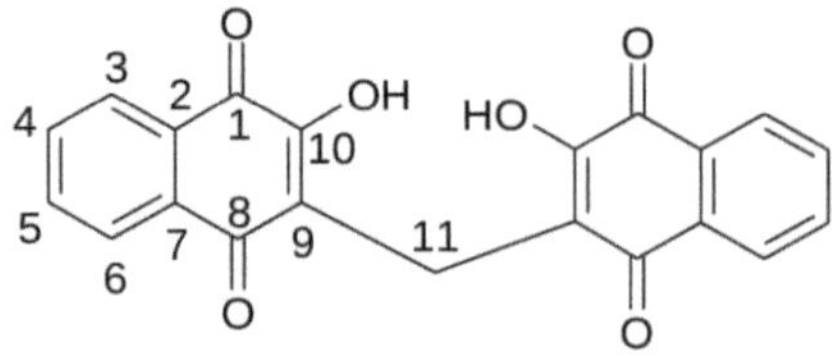

Table 16: **29e** NMR data in *DMSOd6* (δ=ppm, 300 MHz) 1H

|    | $^1H$ | $^{13}C$ |
|----|-------|----------|
| 1  | - | 183,64 |
| 2  | - | 133,15 |
| 3  | 7,94 (dd, 2H, $J_{3,4}$=1,5 Hz) | 131,99 |
| 4  | 7,78 (dd, 2H, $J_{4,5}$=1,2 e $J_{4,3}$=1,5 Hz) | 125,65 |
| 5  | 7,78 (dd, 2H, $J_{5,4}$=1,2 e $J_{5,6}$=1,5 Hz) | 125,93 |
| 6  | 7,96 (dd, 2H, $J_{6,5}$=1,5 Hz) | 129,88 |
| 7  | - | 134,55 |
| 8  | - | 180,78 |
| 9  | - | 122,01 |
| 10 | - | 155,08 |
| 11 | 3,75 (s, 2H) | 17,94 |

## 5.5. Synthesis of 29f: 2,2'-(para-nitrophenylmethylene)bis[3-hydroxy-1,4-naphthalenedione]

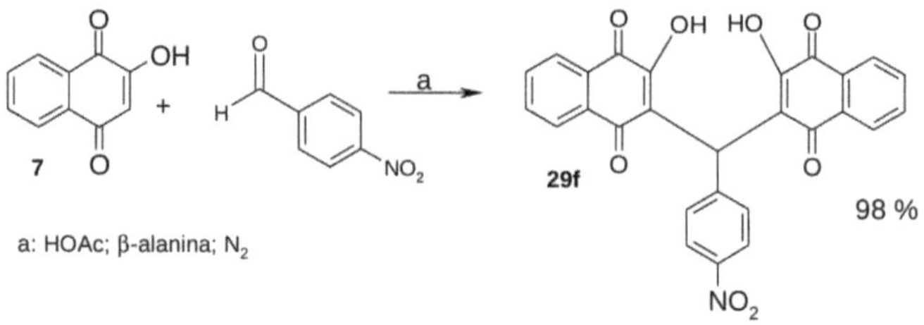

Scheme 39: Condensation with 4-nitro-benzaldehyde

The synthesis of (**29f**) was carried out using the aldehyde known as 4-nitro-benzaldehyde. Isolation was carried out by adding distilled water to the flask and filtering the precipitate, which was washed abundantly with distilled water to remove excess HOAc and 4-nitro-benzaldehyde. The solvent was recrystallized in ethanol, giving red crystals with a yield of 98 % and a melting point showing decomposition at 257 °C. The spectrum of [13] C (DMSO d6, 75 MHz, p. 120) shows 15 types of carbon. The displacement of the aliphatic CH which joins the two molecules at δ 33.62 ppm, the carbon of the aromatic part which is directly linked to the nitro group at δ 150.62 ppm and the carbon *towards* it at δ 145.30 ppm, two peaks referring to carbonyls at δ 183.65 and 182.48 ppm. The[1] H spectrum (*DMSOd6*, 500 MHz, p. 121) shows a singlet at δ 6.81 ppm referring to methinic CH. The IR spectrum (p. 119) shows the absorptions of the hydroxyl groups with the center of the band at 3443 cm[-1] , the axial deformation band of CH sp$^3$ at 2924 cm$^{-1}$ .

Table 3: **29f** NMR data in DMSO-d6 (δ =ppm, 500 MHz)

| | $^1$H* | $^{13}$C** |
|---|---|---|
| 1 | - | 183,65 |
| 2 | - | 133,18 |
| 3 | 7,80 (dt, 2H, $J_{3,5}$=2,0 e $J_{3,4}$=7,0 Hz) | 125,37 |
| 4 | 8,15 (dt, 2H, $J_{4,5}$=2,0; $J_{4,3}$=7,0 e $J_{4,6}$=1,0 Hz) | 132,20 |
| 5 | 7,88 (dd, 2H, $J_{5,4}$=2,0 e $J_{5,6}$=8,0 Hz) | 125,93 |
| 6 | 8,10 (dd, 2H, $J_{6,4}$=1,0 e $J_{6,5}$=8,0 Hz) | 130,97 |
| 7 | - | 134,06 |
| 8 | - | 182,48 |
| 9 | - | 121,66 |
| 10 | - | 164,56 |
| 11 | 6,90 (s, 1H) | 33,66 |
| 12 | - | 145,30 |
| 13 | 8,01 (dd, 2H, $J_{13,14}$=7,0 Hz) | 128,20 |
| 14 | 7,53 (dt, 2H, $J_{14,13}$= 7,0 Hz) | 123,15 |
| 15 | - | 150,62 |

* analysis on a 500 MHz device; ** analysis on a 300 MHz device

## 5.8 Synthesis of 29g: 2,2'-(parahydroxyphenylmethylene)bis[3-hydroxy-1,4-naphthalenedione]

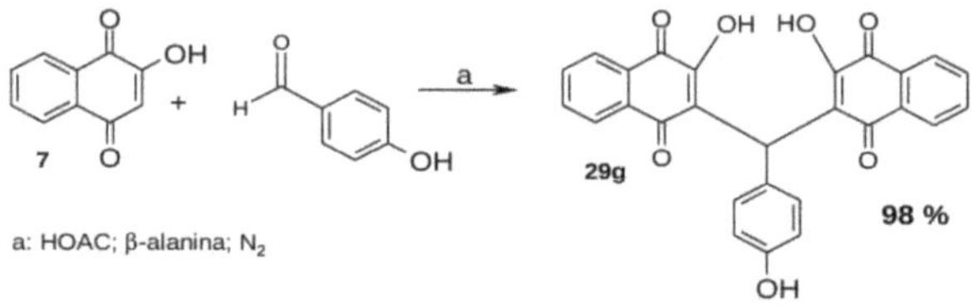

Diagram 40: Condensation with 4-hydroxy-benzaldehyde

The isolation was carried out by adding distilled water to the flask and filtering the precipitate, which was washed abundantly with distilled water to remove the excess HOAc and 4-hydroxy-benzaldehyde. The liquid was recrystallized in ethanol, giving dark red crystals with a yield of 83 % and a melting point of 134-135 °C. The spectrum of[13] C (*DMSOd6*, 75 MHz, p. 123) shows the displacement of 15 types of carbon, with the displacement of the CH joining the two molecules at δ 37.21 ppm, the C-OH of the aromatic part at δ 155.90 ppm, the C *to* the east at δ 155.32 ppm, the carbonyls at δ 183.75 and 181.31 ppm. The spectrum of[1] H (*DMSOd6*, 300 MHz, p. 124) shows a singlet referring to methinic CH at δ 5.89 ppm. The IR spectrum (p. 126) shows the absorption of the phenolic hydroxyl masking the absorptions of the enolic hydroxyls at 3352 cm$^{-1}$ .

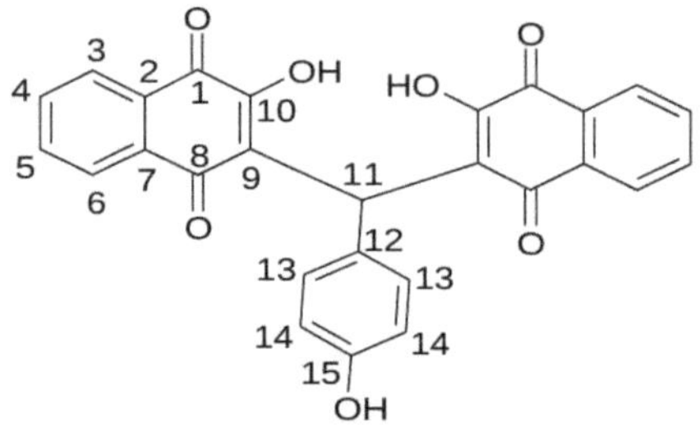

Table 18: NMR data of **29g** in DMSO-d6 (□□=ppm, 300 MHz)

| | ¹H | ¹³C |
| --- | --- | --- |

| 1 | - | 183,85 |
|---|---|---|
| 2 | - | 130,75 |
| 3 | 7,03 (d, 2H, $J_{3,4}$=8,0 Hz) | 125,61 |
| 4 | 7,97 (dt, 2H, $J_{4,3}$=8,0; $J_{4,6}$=0,6 e $J_{4,5}$=6,6 Hz) | 132,20 |
| 5 | 7,82 (dt, 2H, $J_{5,6}$=7,2 e $J_{5,4}$=6,6 Hz) | 129,27 |
| 6 | 7,91 (dd, 2H, $J_{6,5}$=7,2 e $J_{6,4}$=0,6 Hz) | 129,83 |
| 7 | - | 134,71 |
| 8 | - | 181,31 |
| 9 | - | 123,81 |
| 10 | - | 155,32 |
| 11 | 5,89 (s, 1H) | 37,21 |
| 12 | - | 13,16 |
| 13 | 7,75 (t, 2H, $J_{13,14}$= 7,5 Hz) | 126,80 |
| 14 | 6,60 (d, 2H, $J_{14,13}$=7,5 Hz) | 114,55 |
| 15 | - | 155,90 |

## 5.9 Synthesis of 29h: 2,2'-(para-fluorophenylmethylene)bis[3-hydroxy-1,4-naphthalenedione]

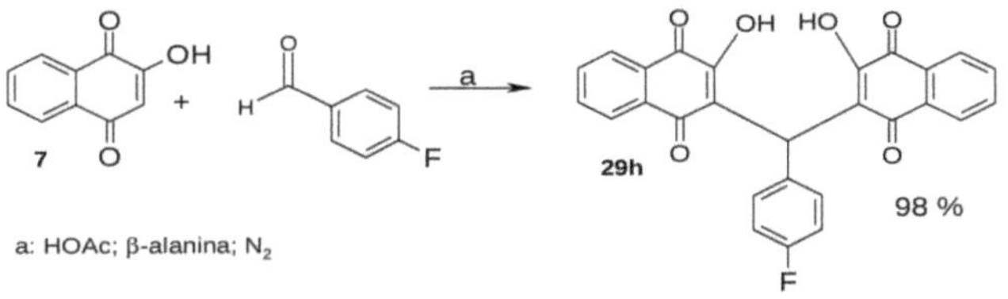

Diagram 41: Condensation with 4-fluor-benzaldehyde

Isolation was carried out by adding distilled water to the flask and filtering the precipitate, which was washed abundantly with distilled water to remove excess HOAc and 4-fluor-benzaldehyde. The solid was obtained in 98% yield and is yellowish in color, with a melting point of 136-137 °C. The IR spectrum (p. 127) shows the absorption of the hydroxyl groups as a strong, broad band at 3414 cm-1. Due to the possibility of coupling between the carbon and the fluorine atom, the 13C NMR spectrum (*DMSOd6*, 125 MHz, p. 128) shows coupling peaks between the fluorine and the carbon atoms of the aromatic. The 1H NMR spectrum (*DMSOd6*, 500 MHz, p. 129) shows the singlet at 6.11 ppm referring to the CH of carbon 11 and the other signals are all aromatic.

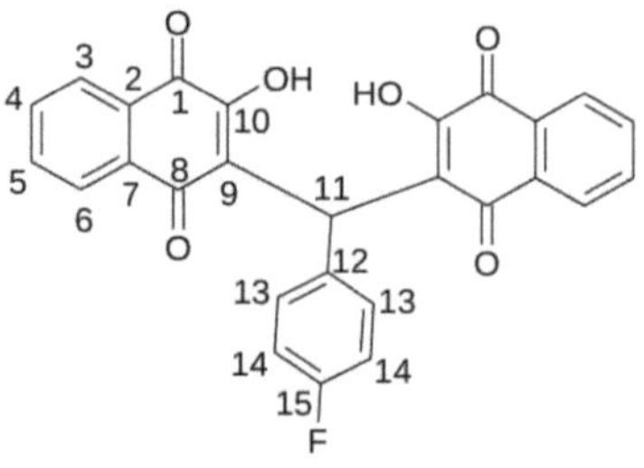

Table 19: **29h** NMR data in DMSO-d6 ($\delta$ =ppm, 500 MHz)

| | [1]H | [13]C |
|---|---|---|
| 1 | - | 183,52 |
| 2 | - | 132,26 |
| 3 | 7,83 (dt, 2H, $J_{3,5}$=1,5 e $J_{3,4}$=7,5 Hz) | 125,58 |
| 4 | 8,10 (dd, 2H, $J_{4,6}$=1,0 e $J_{4,6\,u\,4,3}$=7,5 Hz) | 126,06 |
| 5 | 7,93 (dt, 2H, $J_{5,3}$=1,5; $J_{5,4\,u\,5,6}$=7,5 Hz) | 133,09 |
| 6 | 8,04 (dd, 2H, $J_{6,4}$=1,0 e $J_{6,5}$=7,5 Hz) | 129,97 |
| 7 | - | 134,66 |
| 8 | - | 181,29 |
| 9 | - | 156,44 |
| 10 | - | 123,07 |
| 11 | 6,11 (s, 1H) | 37,10 |
| 12 | - | 136,95 |
| 13 | 7,37 (t, 2H, $J_{13,14}$=9,0 Hz) | 129,93 |
| 14 | 7,11 (t, 2H, $J_{14,13}$=9,0 Hz) | 114,29 |
| 15 | - | 161,48 |

## 5.10. Synthesis of 29i: 2,2'-(2-ethylbutylidene)bis[3-hydroxy-1,4-naphthalenedione]

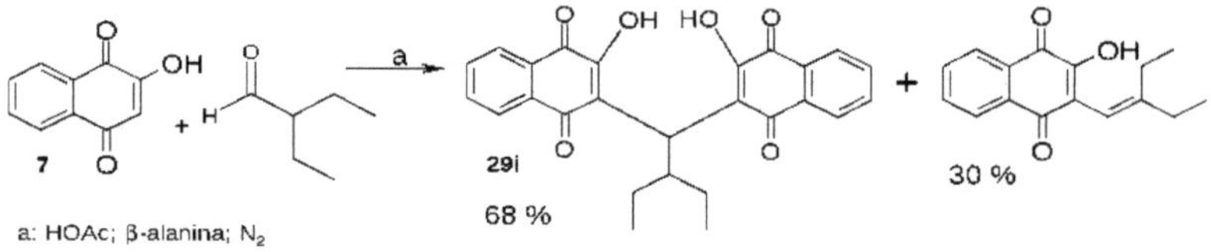

Scheme 42: Condensation with 2-ethyl-butyraldehyde

Isolation was carried out by adding distilled water to the flask and filtering the precipitate, which was washed abundantly with distilled water to remove excess HOAc and 2-ethyl-butyraldehyde. The solvent was obtained in 68 % yield and is orange in color, with a decomposition point of 183 °C. The NMR spectrum[1] H (DMSO *d6*, 500 MHz, p. 130), shows a double triplet at $\delta$ 0.83 ppm, a septet at $\delta$ 1.29 ppm, a duplet referring to the CH joining the two molecules at $\delta$ 5.25 ppm which can be better seen by the expansion shown on p 151 and 152. The spectrum of[13] C (*DMSOd6*, 75 MHz, p. 133) shows the appearance of 14 types of carbon. The difference observed in the[1] H spectrum between the two (-CH2-) in position 13 can be explained by looking at the chemical environment in which each one is found, through the Newman projection of this compound, Figure 12:

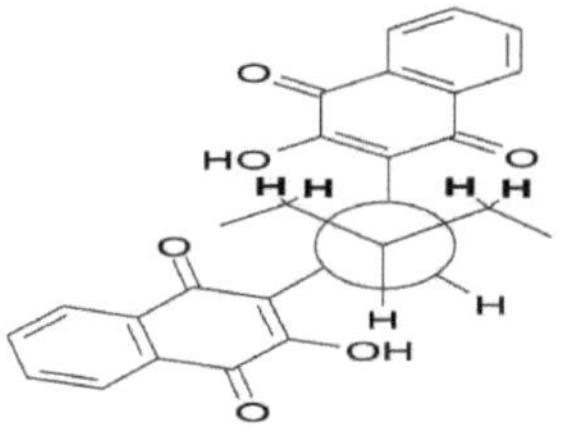

Figure 12: Newman projection of 29i

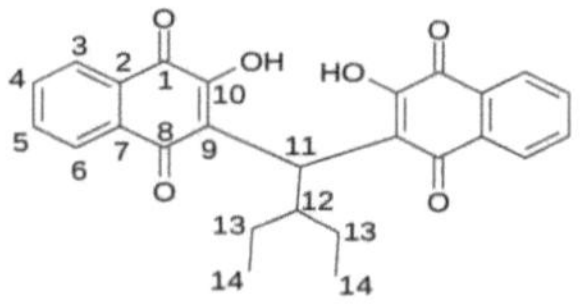

Table 20: **29i** NMR data in *DMSOd6* (δ =ppm, 500 MHz)

| | $^1$H | $^{13}$C |
|---|---|---|
| 1 | - | 183,49 |
| 2 | - | 132,35 |
| 3 | 8,03 (dt, 2H, $J_{3,4}$=7,0 Hz) | 134,16 |
| 4 | 7,74 (dd, 2H, $J_{4,3}$=7,0 e $J_{4,5}$=7,5 Hz) | 125,19 |
| 5 | 7,82 (dt, 2H, $J_{5,4}$=7,5 e $J_{5,6}$=7,5 Hz) | 125,93 |
| 6 | 7,94 (d, 2H, $J_{6,5}$=7,5 Hz) | 130,25 |
| 7 | - | 132,80 |
| 8 | - | 182,71 |
| 9 | - | 123,40 |
| 10 | - | 160,57 |
| 11 | 5,34 (d, 1H, $J_{12,11}$=12 Hz) | 35,61 |
| 12 | 1,43 (m, 1H) | 33,73 |
| 13 | 1,28 (q, 4H, $J_{13,14}$= 7,5 Hz) | 22,03 |
| 14 | 0,83 (t, 6H, $J_{14,13}$=7,5 Hz) | 10,01 |

## 5.11. Synthesis of 29j: 2,2'-(ethylidene)bis[3-hydroxy-1,4-naphthalenedione]

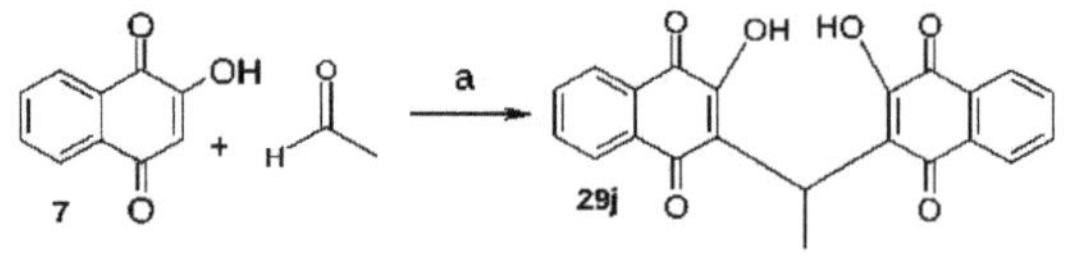

Diagram 43: Condensation with acetaldehyde

Isolation was carried out by adding distilled water to the flask and filtering the precipitate, which was washed abundantly with distilled water to remove excess HOAc and acetaldehyde. A dark yellow colored liquid was obtained in 72% yield, with a melting point in the 175-176 °C range. The[1] H spectrum (DMSO $d6$, 300 MHz, p. 134) shows the three basic signals of the dimer structure, a duplet at δ 1.60 ppm and a double duplet with a center at δ 4.61 ppm. Analysis of the[13] C spectrum (DMSO-d6, 75 MHz, p. 135) shows that these two aliphatic carbons correspond to the displacement of δ 28.51 and 17.23 ppm. A product of lower polarity was observed by CCDA, but its characterization was not carried out.

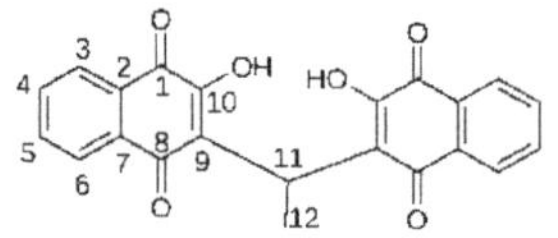

Table 21: **29j** NMR data in *DMSOd6* (δ =ppm, 300 MHz)

|  | [1]H | [13]C |
|---|---|---|
| 1 | - | 181,30 |
| 2 | - | 132,16 |
| 3 | 7,72 (dt, 2H, $J_{3,4}$=6,3 Hz) | 134,61 |
| 4 | 7,93 (dt, 2H, $J_{4,5}$=7,5 e $J_{4,6}$=1,5 Hz) | 125,76 |
| 5 | 7,83 (dt, 2H, $J_{5,4}$=7,5; $J_{5,6}$=7,5 e $J_{5,3}$=1,8 Hz) | 125,97 |
| 6 | 7,91 (dt, 2H, $J_{6,5}$=7,5 Hz) | 129,61 |
| 7 | - | 133,02 |
| 8 | - | 183,84 |
| 9 | - | 125,49 |
| 10 | - | 155,07 |
| 11 | 4,61 (q, 1H, $J_{11,12}$=7,2 Hz) | 28,51 |
| 12 | 1,61 (d, 3H, $J_{12,11}$=7,2 Hz) | 17,23 |

The carbon spectrum shows the appearance of 12 types of peaks, demonstrating molecular symmetry. It is not possible to distinguish from this spectrum some carbons that have very close displacements. To achieve these definitions with greater certainty, it would be necessary to have a spectrum that distinguishes between CH and C quaternary carbons. Table 22 shows the elemental analysis data for the dimers:

Table 22: Elemental analysis of dimers

| Composto | % (C)t/e | % (H)t/e | % (O)t/e |
|---|---|---|---|
| 29e | 70,00/68,79 | 3,36/3,05 | 26,64/28,14 |
| 29j | 70,59/62,10 | 3,77/3,23 | 25,64/34,65 |
| 29a | 71,64/56,77 | 4,51/3,58 | 23,86/39,64 |
| 29 | 71,64/65,68 | 4,51/3,81 | 23,86/30,49 |
| 29i | 72,55/44,34 | 5,15/2,99 | 22,30/52,65 |
| 29b | 72,55/61,76 | 5,15/4,00 | 22,30/34,23 |
| 29c | 74,31/64,49 | 3,70/3,60 | 22,00/31,90 |
| 29g | 71,68/55,89 | 3,56/2,95 | 24,76/41,14 |
| 29d | 72.10/72,36 | 3,89/3,91 | 24.01/23,71 |
| 29h | 68,94/69,92 | 3,21/3,09 | 23,81/21,00 |

Due to various experimental factors, it was only possible to complete the characterization of 29d using this technique

# 6.   CONCLUSION

Reactions using DDQ in an apolar solvent such as hexane, as proposed in the initial methodology, were unsuccessful, with many by-products being formed and the desired product (**11**) having a very low yield. It was noticeable that the DDQ did not solubilize completely. When the solvent was changed to methanol, the reaction gave a five-membered cyclic product with an addition of the solvent itself to the molecule.

The search for the synthesis of amino-naphthoquinones led to the production of a molecule never before seen in the literature (**4b**). This molecule was analyzed by X-ray diffraction and its data was submitted to the *Journal Molecular Structure*.

A process for synthesizing nor-lapachol was developed with a yield of 96% and purification was carried out only by extraction in a basic medium.

When nor-lapachol was synthesized using the Kopanski method, dimeric molecules were formed. A methodology was developed to synthesize them in good yields, and the products were characterized by NMR spectroscopy, infrared spectroscopy and elemental analysis.

The products were obtained in good yields from the condensation of lausone with aliphatic and aromatic aldehydes.

Changing the benzene solvent to acetic acid resulted in better yields and shorter reaction times.

The products were characterized using conventional techniques.

# 7. EXPERIMENTAL

## 7.1. Synthesis of 29: 2,2'-(2-methylpropylidene)bis[3-hydroxy-1,4-naphthalenedione]

For the preparation of dimer (**29**), the methodology in which benzene is the reaction solvent was used, weighing 10 mmol of 2-hydroxy-1,4-naphthoquinone (**7**) (1.74 g), 20 mmol of isobutyraldehyde (1.4688 g), 0.15 g of β-alanine and 0.25 g of HOAc. The reaction was carried out in a nitrogen atmosphere and kept under heat of 50 °C for 24 hours. Isolation was carried out by removing the benzene under a rotaevaporator at low pressure, after which a chromatographic column was made using silica gel as the fixed phase and hexane/ethyl acetate (9:1) to remove the part of the corresponding alkene that was formed. An orange-colored product with a decomposition point of 185-186 °C was isolated. The yield was 58 %. In the analysis of the infrared spectrum, it was possible to observe an absorption band in the region of compounds with hydroxyls, the absorption was broad and strong at 3452 cm$^{-1}$ , characteristic of an asymmetric stretching of the -O-H bond, a very characteristic absorption of the -C-H sp bond$^{3}$ referring to methyl groups at 2952 cm$^{1}$ the abosorption of the carbonyl part, as expected, appeared as two strong bands referring to aromatic carbonyls at 1670 cm$^{1}$ and conjugated carbonyl at 1627 cm$^{-1}$ .

**IV (KBr)** v max 3452, 2952, 1670, 1627, 1597, 1570, 1458, 1361, 1280, 1222, 1107, 968, 729, cm$^{-1}$ .

**NMR[1] H (*DMSOd6*, 300 MHz, ppm)** 7.90 (d, 2H, *J3.4=7* Hz); 7.62 (dt, 2H, *J4.3=7.0* and *J4.5=5.5* Hz); 7.71 (dt, 2H, *J5.4=7.2* and *J5.6=6.6* Hz); 7.83 (d, 2H, *J6.5=7.5* Hz); 4.93 (d, 1H, *J11.12=10.8* Hz); 2.98 (m, 1H, *J12.13=6.3* and *J12.11=10.8* Hz); 0.78 (d, 6H, *J13.12=6.3* Hz).

**NMR[13] C (*DMSOd6*, 75 MHz, ppm)** 183.14; 133.14; 131.82; 125.69; 124.89; 130.79; 133.71; 183.09; 163.44; 123.48; 36.39; 24.16; 21.65.

## 7.2. Synthesis of 29a: 2,2'-(2-butylidene)bis[3-hydroxy-1,4-naphthalenedione]

For the synthesis of dimer (**29a**), the methodology in which benzene is the reaction solvent was used, weighing 10 mmol of 2-hydroxy-1,4-naphthoquinone (**7**) (1.74 g), 20 mmol of *n-butyraldehyde* (1.4688 g), 0.15 g of β-alanine and 0.25 g of HOAc. The reaction is carried out in a nitrogen atmosphere and kept under heat of 50 °C for 24 hours. Isolation is carried out by removing the benzene under a rotaevaporator at low pressure, after which a chromatographic column is made using silica gel as the fixed phase and hexane/ethyl acetate (9:1) to remove the part of the corresponding alkene that is formed. An orange-colored product with a decomposition point between 206-207 °C was isolated, with a yield of 36 %.

**IV(KBr)**ν max 3452, 2958, 2927, 1666, 1627, 1597, 1570, 1458, 1361, 1280, 1261, 1122, 964, 732 cm$^{-1}$ .

**NMR$^1$ H (*DMSOd6*, 300 MHz, ppm)** 7.91 (d, 2H, *J3.4=7.0* Hz); 7.62 (dt, 2H, *J4.3=7.0* and *J4.5=7.5* Hz); 7.72 (dt, 2H, *J5.4=7.5* and *J5.6=7.5* Hz); 7.83 (t, 2H, *J6.5=7.5* Hz); 5.31 (t, 1H, *J11.12=8.1* Hz); 1.54 (dd, 2H, *J12.11=8.1* and *J12.13=7.5* Hz); 1.15 (dt, 2H, *J13.12=7.5* and *J13.14=7.2* Hz); 0.83 (t, 2H, *J14.13=7.2* Hz)

**NMR$^{13}$ C (*DMSOd6*, 75 MHz, ppm)** 184.12; 133.24; 125.05; 131.92; 125.74; 130.72; 133.90; 182.77; 163.82; 123.54; 38.66; 30.87; 21.09; 14.11.

### 7.3. **Synthesis of 29b: 2,2'-(hexylidene)bis[3-hydroxy-1,4-naphthalenedione]**

The synthesis of (**29b**) was carried out using acetic acid as the solvent. We weighed 20 mmol of hexanaldehyde (2.0 g), 1.74 g of lausone (10 mmol), 0.15 g of β-alanine and approximately 25 mL of HOAc. The system was kept at 50 °C in a nitrogen atmosphere for 35 minutes. Isolation is carried out by adding distilled water to the flask and filtering the precipitate, which is washed abundantly with distilled water to remove excess HOAc and hexanaldehyde. The final product was an orange-colored solid, obtained in a yield of 69% and with a decomposition point of 256 °C. Analysis of the spectrum of$^1$ H carried out in (*DMSOd6*, 300 MHz) shows a triplet at 0.77 ppm referring to the terminal methyl of the aliphatic chain with the neighboring CH2, a triplet with a shift of 5.2 ppm referring to the CH that joins the two molecules. IR absorptions show two absorption bands at 2927 and 2854 cm$^1$ which are due to the asymmetric stretching of CH sp$^3$ . The same absorptions of the carbonyls observed for (**29a**) are observed in (**29b**), which are the bands at 1666 and 1627 cm .$^{-1}$

**IV (KBr)** ν max 3433, 2927, 2854, 1666, 1627, 1585, 1566, 1458, 1365, 1280, 1230, 952, 736 cm$^{-1}$ .

**NMR$^1$ H (*DMSOd6*, 300 MHz, ppm)** 7.92 (dd, 2H, *J3.4=7.5 and J3.5=0.9* Hz); 7.63 (dt, 2H, *J4.3=7.5; J4.5= 7.2* and *J4.6=1.2* Hz); 7.72 (dt, 2H, *J5.4=7.2; J5.6=7.5* and *J5.3=0.9* Hz); 7.85 (dd, 2H, *J6.5=7.5 and J6.4=1.2* Hz); 5.18 (t, 1H, *J11.12=8 Hz); 1.*99 (m, 2H); 1.18 (m, 2H); 1.18 (m, 2H); 1.18 (m, 2H); 0.77 (t, 3H, *J16.15=6.9* Hz).

**NMR$^{13}$ C (*DMSOd6*, 75 MHz, ppm)** 183.64; 133.10; 125.04; 131.98; 125.73; 130.56; 133.90; 182.87; 162.70; 123.68; 27.61; 31.37; 28.76; 22.08; 29.10; 13.97.

### 7.4. Synthesis of 29c: 2,2'-(phenylmethylene)bis[3-hydroxy-1,4-naphthalenedione]

The synthesis of (**29c**) was carried out using acetic acid as the solvent. We weighed 20 mmol of benzaldehyde (1.92 g), 1.74 g of lausone (10 mmol), 0.15 g of β-alanine and approximately 25 mL of HOAc. The system was kept at 50 °C in a nitrogen atmosphere for 4 hours. Isolation is carried out by adding distilled water to the flask and filtering the precipitate, which is washed thoroughly with distilled water to remove excess HOAc and benzaldehyde. The final product is a yellow-colored solid, obtained in 93 % yield and with a decomposition point of 273 °C.

**IV (KBr)** v max 3450, 1674, 1597, 1570, 1361, 1284, 1222, 1111, 1056, 729 cm$^{-1}$

**NMR$^1$ H (*DMSOd6*, 300 MHz, ppm)** 7.72 (d, 2H, *J3.4=7.5* Hz); 7.76 (t, 2H, *J4.3* and *4.5=7.5* Hz); 7.67 (t, 2H, *J5.4* and *5.6= 7.5* Hz); 7.89 (d, 2H, *J6.5=7.5* Hz); 6.70 (s, 1H).

**NMR$^{13}$ C (*DMSOd6*, 75 MHz, ppm)** 183.84; 133.27; 124.90; 132.05; 125.84; 130.89; 133.98; 182.63; 164.27; 122.84; 33.08; 141.27; 125.22; 127.83; 126.85.

### 7.5. Synthesis of 29d: 2,2'-(para-methoxyphenylmethylene)bis[3-hydroxy-1,4-naphthalenedione]

This reaction was carried out using acetic acid as a solvent. To do this, 20 mmol of anisic aldehyde (2.7744 g), 10 mmol of lausone (1.74 g), 0.15 g of β-alanine and approximately 25 mL of glacial acetic acid were weighed. The system was kept at a temperature of 50 °C in a nitrogen atmosphere for 4 hours. Isolation is carried out by adding distilled water to the flask and filtering the precipitate, which is washed thoroughly with distilled water to remove excess HOAc and anisic aldehyde. The final product is a yellow-colored solid, obtained in 94% yield and with a melting point in the range of 224-225 °C.

**IV(KBr)**v max3394, 1666, 1639, 1593, 1512, 1458, 1361, 1338, 1276, 1261, 1238, 1045, 1018, 721 cm$^{-1}$

**NMR$^1$ H (*DMSOd6*, 300 MHz, ppm)** 7.98 (dd, 2H, *J3.4=7.0* and *J3.5=1.2* Hz); 7.77 (dt, 2H, *J4.3=7.0*; *J4.6=1.2* and *J4.5=7.5* Hz); 7.77 (dt, *J5.3=1.2*; *J5.4=7.5* and *J5.6=7.2* Hz); 7.98 (dd, *J6.5=7.2* and *J6.4=1.2* Hz); 5.94 (s, 1H); 6.78 (dd, 2H, *J13.14=8.7* Hz); 7.18 (dd, 2H, *J14.13=8.7* Hz); 3.69 (s, 3H).

**NMR$^{13}$ C (*DMSOd6*, 75 MHz, ppm)** 183.74; 132.70; 125.64; 132.21; 126.11; 129.85; 133.16; 181.28; 123.63; 155.90; 37.31; 134.73; 129.37; 113.10; 157.34; 54.94.

### 7.6. Synthesis of 29e: 2,2'-(methylene)bis[3-hydroxy-1,4-naphthalenedione]

Because formaldehyde has a large amount of water in its composition, the solvent known as paraformaldehyde was used to carry out this reaction. We used 20 mmol of the aldehyde (612 mg), 10 mmol of lausone (1.74 g), 0.15 g of β-alanine and approximately 25 mL of HOAc. The system was kept at a temperature of 50 °C in a nitrogen atmosphere for 6 hours. Isolation is carried out by adding distilled water to the flask and filtering the precipitate, which is washed thoroughly with distilled water to remove excess HOAc and paraformaldehyde. The final product is a yellow-colored liquid, obtained in a yield of 75% and with a degradation point of 251 °C. The literature value is 249251 dec.

**IV(KBr)**ν max3452,3070, 1678, 1610, 1573, 1458, 1350, 1323, 1265, 1215, 975, 937, 771, 736,466 cm$^{-1}$

**NMR$^1$ H (*DMSOd6*, 300 MHz, ppm)** 7.94 (dd, 2H, *J3.4=1.5* Hz); 7.78 (dd, 2H, *J4.5=1.2* and *J4.3=1.5* Hz); 7.78 (dd, 2H, *J5.4=1.2* and *J5.6=1.5* Hz); 7.96 (dd, 2H, *J6.5=1.5* Hz); 3.75 (s, 2H).

**NMR$^{13}$ C (*DMSOd6*, 75 MHz, ppm)** 183.64; 133.15; 125.65; 131.99; 125.93; 129.88; 134.55; 180.78; 122.01; 155.08; 17.94.

### 7.7. **Synthesis of 29f: 2,2'-(para-nitrophenylmethylene)bis[3-hydroxy-1,4-naphthalenedione]**

The synthesis of (**2f**) was carried out using the aldehyde known as 4-nitro-benzaldehyde, using 2 mmol of this aldehyde (302 mg), 1 mmol of lausone (0.174 g), 0.05 g of β-alanine and 10 mL of glacial acetic acid. The system was kept at a temperature of 50 °C in a nitrogen atmosphere for 1 hour and 15 minutes. Isolation was carried out by adding distilled water to the flask and filtering the precipitate, which was washed abundantly with distilled water to remove excess HOAc and 4-nitro-benzaldehyde. The liquid was recrystallized in ethanol, giving red crystals with a yield of 98 % and a decomposition point of 257 °C.

**IV (KBr)** ν max3433, 2924, 1670, 1597, 1570, 1512, 1350, 1280, 1111, 732 cm$^{-1}$

**NMR$^1$ H (*DMSOd6*, 500 MHz, ppm)** 7.80 (dt, 2H, *J3.5=2.0* and *J3.4=7.0* Hz); 8.15 (dt, 2H, *J4.5=2.0*; *J4.3=7.0* and *J4.6=1.0* Hz); 7.88 (dd, 2H, *J5.4=2.0* and *J5.6=8.0* Hz); 8.10 (dd, 2H, *J6.4=1.0* and J6.5=8.0 Hz); 6.90 (s, 1H); 8.01 (dd, 2H, *J13.14=7.0* Hz); 7.53 (dt, 2H, *J14.13= 7.0* Hz).

**NMR$^{13}$ C (*DMSOd6*, 300 MHz, ppm)** 183.65; 133.18; 125.37; 132.20; 125.93; 130.97; 134.06; 182.48; 121.66; 164.56; 33.66; 145.30; 128.20; 123.15; 150.62.

7.8. **Synthesis of 29g: 2,2'-(parahydroxyphenylmethylene)bis[3-hydroxy-1,4-naphthalenedione]**

To synthesize (**29g**), 4-hydroxy-benzaldehyde was used in the proportion of 20 mmol (2.44 g), 10 mmol of lausone (1.74 g), 0.15 g of β-alanine and approximately 25 mL of glacial acetic acid. The system was kept at a temperature of 50 °C in a nitrogen atmosphere for 55 minutes. Isolation was carried out by adding distilled water to the flask and filtering the precipitate, which was washed abundantly with distilled water to remove excess HOAc and 4-hydroxy-benzaldehyde. The solvent was recrystallized in ethanol, giving dark red crystals with a yield of 83% and a melting point in the range 134-135°C.

**IR (KBr)** v max 3352, 1647, 1593, 1512, 1458, 1365, 1276, 1045, 1010, 972, 902, 725 cm$^{-1}$

**NMR$^1$ H (*DMSOd6*, 300 MHz, ppm)** 7.03 (d, 2H, *J3.4*=8.0 Hz); 7.97 (dt, 2H, *J4.3*=8.0; *J4.6*=0.6 and *J4.5*= 6.6 Hz); 7.82 (dt, 2H, *J5.6*=7.2 and *J5.4*=6.6 Hz); 7.91 (dd, 2H, *J6.5*=7.2 and *J6.4*=0.6 Hz); 5.89 (s, 1H); 7.75 (t, 2H, *J13.14*= 7.5 Hz); 6.60 (d, 2H, *J14.13*=7.5 Hz).

**NMR$^{13}$ C (*DMSOd6*, 75 MHz, ppm)** 183.85; 130.75; 125.61; 132.20; 129.27; 129.83; 134.71; 181.31; 123.81; 155.32; 37.21; 13.16; 126.80; 114.55; 155.90.

7.9. **Synthesis of 29h: 2,2'-(para-fluorophenylmethylene)bis[3-hydroxy-1,4-naphthalenedione]**

The dimer (**29h**) was synthesized from 4-fluor-benzaldehyde, using 20 mmol of this aldehyde (2.44 g), 10 mmol of lausone (1.74 g), 0.15 g of β-alanine and approximately 25 mL of glacial acetic acid. The system was kept at a temperature of 50 °C in a nitrogen atmosphere for 1 hour and 15 minutes. Isolation is carried out by adding distilled water to the flask and filtering the precipitate, which is washed abundantly with distilled water to remove excess HOAc and 4-fluor-benzaldehyde. The solvent is obtained in 98% yield and is yellowish in color, with a melting point in the range 136-137°C.

**IV(KBr)**v max3414,3348, 1666, 1625, 1593, 1508, 1458, 1365, 1342, 1276, 1230, 1161, 1041, 833,725 cm$^{-1}$

**NMR$^1$ H (*DMSOd6*, 500 MHz, ppm)** 7.83 (dt, 2H, *J3.5*=1.5 and *J3.4*=7.5 Hz); 8.10 (dd, 2H, *J4.6*=1.0 and *J4.5* and 4.3=7.5 Hz); 7.93 (dt, 2H, *J5.3*=1.5; *J5.4* and 5.6=7.5 Hz); 8.04 (dd, 2H, *J6.4*=1.0 and *J6.5*=7.5 Hz); 6.11 (s, 1H); 7.37 (t, 2H, *J13.14*=9.0 Hz); 7.11 (t, 2H, *J14*.13=9.0 Hz).

**NMR[13] C (*DMSOd6*, 125 MHz, ppm)** 183.52; 132.26; 125.58; 133.09; 126.06; 129.97; 134.66; 181.29; 123.07; 156.44; 37.10; 136.95; 129.93; 114.29; 161.48.

### 7.10. Synthesis of 29i: 2,2'-(2-ethylbutylidene)bis[3-hydroxy-1,4-naphthalenedione]

The synthesis of dimer (**29i**) was proposed with an aliphatic aldehyde, known as 2-ethyl-butyraldehyde, where 20 mmol of the aldehyde in question (2.0 g), 10 mmol of lausone (1.74 g), 0.15 g of β-alanine and approximately 25 mL of glacial acetic acid were used for the reaction. The system was kept at a temperature of 50 °C in a nitrogen atmosphere for 2 hours and 15 minutes. Isolation was carried out by adding distilled water to the flask and filtering the precipitate, which was washed abundantly with distilled water to remove the excess HOAc and 2-ethyl-butyraldehyde. The solid was obtained in a yield of 68 % and is orange in color, with a decomposition point in the range of 183 °C.

**NMR[1] H (*DMSOd6*, 500 MHz, ppm)** 8.03 (dt, 2H, $J3.4{=}7.0$ Hz); 7.74 (dd, 2H, $J4.3{=}7.0$ and $J4.5{=}7.5$ Hz); 7.82 (dt, 2H, $J5.4{=}7.5$ and $J5.6{=}7.5$ Hz); 7.94 (d, 2H, $J6.5{=}7.5$ Hz); 5.34 (d, 1H, $J12.11{=}12$ Hz); 1.43 (m, 1H); 1.28 (q, 4H, $J13.14{=}7.5$ Hz); 0.83 (t, 6H, $J14.13{=}7.5$ Hz).

**NMR[13] C (*DMSOd6*, 125 MHz, ppm)** 183.49; 132.35; 125.19; 134.16; 125.93; 130.25; 132.80; 182.71; 123.40; 160.57; 35.61; 33.73; 22.03; 10.01.

### 7.11. Synthesis of 29j: 2,2'-(ethylidene)bis[3-hydroxy-1,4-naphthalenedione]

This synthesis is made difficult by the fact that acetaldehyde is a liquid with a high vapor pressure, which makes it difficult to collect with a pipette. Even so, it was possible to collect approximately 20 mmol (0.885 g), 1.74 g of lausone, corresponding to 10 mmol, 0.15 g of β-alanine and approximately 25 mL of glacial acetic acid. The system was kept at a temperature of 50 °C in a nitrogen atmosphere and after 5 minutes it was observed that the reaction medium became clear, turning cloudy again after a few more minutes, and the reaction was completed after 1 hour and 15 minutes. Isolation was carried out by adding distilled water to the flask and filtering the precipitate, which was washed thoroughly with distilled water to remove excess HOAc and acetaldehyde. A dark yellow colored liquid was obtained with a yield of 72 % and a melting point in the range of 175-176 °C.

**NMR[1] H (*DMSOd6*, 500 MHz, ppm)** 7.72 (dt, 2H, $J3.4{=}6.3$ Hz); 7.93 (dt, 2H, $J4.5{=}7.5$ and $J4.6{=}1.5$ Hz); 7.83 (dt, 2H, $J5.4{=}7.5$; $J5.6{=}7.5$ and $J5.3{=}1.8$ Hz); 7.91 (dt, 2H, $J6.5{=}7.5$ Hz); 4.61 (q, 1H, $J11.12{=}7.2$ Hz); 1.61 (d, 3H, $J12.11{=}7.2$ Hz).

**NMR[13] C (*DMSOd6*, 125 MHz, ppm)** 183.49; 132.35; 125.19; 134.16; 125.93; 130.25; 132.80; 182.71; 123.40; 160.57; 35.61; 33.73; 22.03; 10.01.

# 8. Biological tests

## 8.1. Tests with the human topoisomerase enzyme II

**1-DNA (0.152 µg/mL) plasmid pB322; 2-DNA+topo II (1Unit.); 3-DNA+topo II+1Unit.)+etoposide**

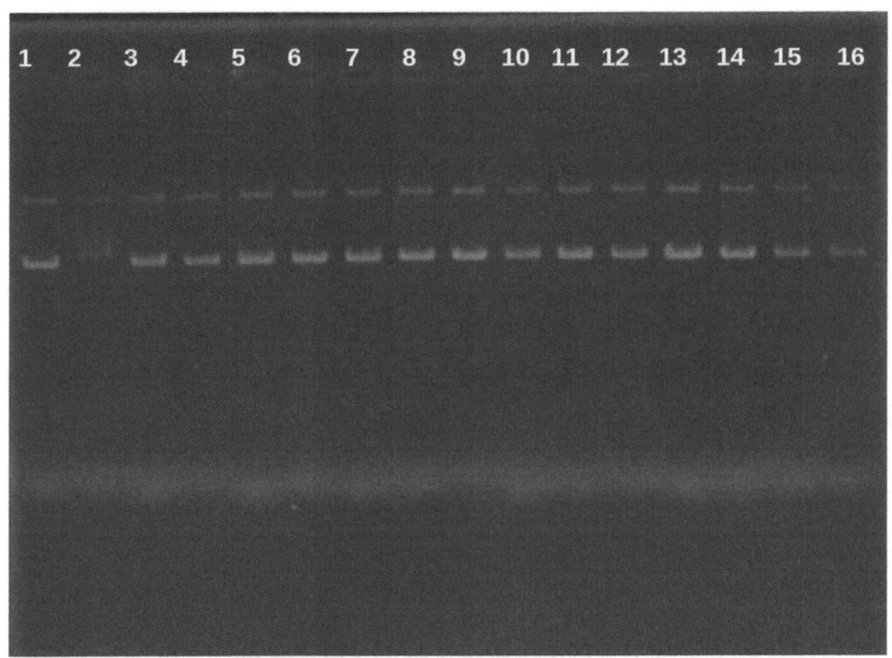

(100µM); 4-DNA+topo II+1Unid.)+Riparin 1 (50µM) 5-DNA+topo II+1Unid.)+29 ; 6-DNA+topo II+1Unid.)+29a (100µM); 7- DNA+topo II+1Unid.)+29b (100µM); 8- DNA+topo II+1Unid.)+ 29c (100µM); 9- DNA+topo II+1Unid.)+ 29d (100µM); 10- DNA+topo II+1Unid.)+ 29e (100µM); 11- DNA+topo II+1Unid.)+ác. 29f (100µM); 12- DNA+topo II+1Unid.)+ 29g (100µM); 13- DNA+topo II+1Unid.)+ 29h (100µM); 14- DNA+topo II+1Unid.)+ 29i (100µM); 15- DNA+topo II+1Unid.)+Waref. (100µM);16- DNA+topo II+1Unid.)+ DMSO 2%

These results show that the substances were active in inhibiting the catalytic action of the human topoisomerase enzyme 11. However, the white test, which corresponds to 16, showed activity. This probably occurred because the solvent was contaminated or its concentration was too high when dissolving the organic sample to be studied.

## 8.2. Toxicity tests against *Artemia salina* Leach

A *screening* was carried out with concentrations of 50 µg/ml of all the dimers, except 29f, and the solution was made with seawater collected at a distance of approximately 50 meters from the bathers.

We weighed out 20 mg of each dimer and added three drops of cremofor®, a detergent to facilitate solubilization, and 2 ml of DMSO. The final volume was 10 ml, topped up with seawater.

We used 10 samples and a blank with 2 ml of DMSO and 8 ml of seawater, from which 125 µl was taken and topped up to 5 ml with saline water. This procedure was carried out for all 10 samples.

The result was that there were no deaths in any of the samples. Now we need to carry out the same experiment with a concentration of 500 µg/ml.

# 9. REFERENCES

Barbosa, T. P.; Camara, C. A.; Silva, T. M. S.; Martins, R. M. Pinto, A. C. and Vargas, M. D. *Bioorg. & Med. Chem.*, 2005, 13, 6464-6469.

Behforouz, M.; Gu, Z; Stelzer, L. S.; Ahmadian, M.; Haddad, J. and Scherschel, J. A. *Tetrahedron. Lett.*, 1997, 38, 2211-2214.

Camara, C. A.; Pinto, A. C.; Torres, J. C.; and Vargas, M. D. Abstracts from the *XVI Meeting of the Brazilian Chemical Society*, Poços de Caldas - Minas Gerais, May 2003, QO 044.

Camara, C. A.; Pinto, A. C.; Vargas, M. D. and Zuckermann, S., J. *Tetrahedron*, 2002, 58, 61356140.

Camara, C. A.; Rosa, M. A.; Pinto, A. C. and Vargas, M. D. *Tetrahedron*, 2001, 57, 9569-9574.

Catti, F.; Kiuru, P. S.; Slawin, A. M. Z. and Westwood, N. J. *Tetrahedron*, 2008, 64, 9561-9566.

Cunha, A. S.; Lima, E. L. S.; Pinto, A. C.; Souza, A. E.; Echevarria, A.; Camara, C. A.; Vargas M. D. and. Torres, J. C. *J. Braz. Chem. Soc*, 2006, 17, 439-442.

Dudley, K. H. and Chiang, R. W. *J. Org. Chem.* 1969, 34, 120-126.

Farouk, A.; Giles, R. G. F.; Green, I. R. and Pearce, R. *Synth. Comm.*, 2004, 1247-1258p.

Ghera, E.; David, Y. B.and Rapoport, H. *J. Org. Chem.*, 1981, 46, 2059-2065.

Gontijo, B. and Carvalho, M. L. R. *Rev. Soc. Bras. Med. Trop.*, 2003, 36, 71-80.

Hansson, S.; Heumann, A.; Rein, T. and Aakermark, B. *J. Org. Chem.*, 1990, 55, 975-84.

Hayashi, T.; Smith, F. T. and Lee, K. *J. Med. Chem*, 1987, 30, 2005-2008.

Heck, R. F. *Palladium Reagents in Organic Synteses*, Academic Press, New York,1985, 74p.

Hooker, S. C. *J. Am. Chem. Soc.*, 1936, 58, 1163-1167.

Hsieh, T.; Chang, F.; Chia, Y.; Chen, C.; Chiu, H. and Wu, Y. *J. Nat. Prod.* 2001, 64, 616-619.

Hudlicky. M. *Oxidations in Organic Chemistry*, ACS Monographies series, NY, 1990.

Jones, G. *The Quinolines*. London: Wiley-Interscience, 1977, p. 93-318.

Kar, A. and Nargade, N.P. *Tetrahedron,* 2003, 59, 2991-2998.

Kirchlechner, R.; Casutt, M.; Heywang, U. and Schwarz, M. W. *Synthesis*, 1994, 247.

Kolodina, E. A.; Lebedeva, N. I. and Shvartsberg, M. S. *Rus. Chem. Bull.*, 2007, 56, 2466-2470.

Kopanski, L.; Karbach, D.; Selbifsehka, G. and Steglich, W. *Annal. Chem.* 1987, 53, 793-796.

Koyamad, J.; Tagahaya, K. and Nishinob, H. *Cancer Lett.*, 1997. 113, 47-53.

Kumar, S.; Malachowski, W. P.; DuHadaway, J. B.; LaLonde, J. M.; Carroll, P. J.; Jaller, D.;

Metz, R.; Prendergast, G. C. and Muller, A. J. *J. Med. Chem. 2008,* 51, 1706-1718.

Larock, R. C. and Kuo, M., *Tetrahedron Lett.,* 1991, 32, 569-572.

Lee, K.; Turnbull, P. and Moore, H. W. *J. Org. Chem.*, 1995, 60, 461-464.

Lee, Y. R. and Lee, W. K., *Synthetic Com.* 2004, 34, 4537-4543.

Lima, N. M. F.; Correia, C. S.; Ferraz, P. A. L.; Pinto, A. V.; Pinto, M. C. R. F.; Santana, A. E. G. and Goulart, M. O. F., *J. Braz. Chem. Soc.*, 2002, 13, 822-829.

Liu, D.; Wikstro, H. K.; Dijkstra, M. D. D.; de Vries, J. V. and Bastiaan, J. V. *J. Med. Chem.* 2006, 49, 1494 1498.

Loredana, B. and Palitti, F. *Gen. Mol. Biol.*, 2000, 23, 1065-1069.

Mahindra, T. M. and Vithal, M. H. *Bioorg. Med. Chem.*, 2002, 10, 1483-1497.

Mahindra, T. M. and Vithal, M. K. *J. Comp. Aid Molecular. Des.*, 2001, 15, 961-978.

McKinney, J. D. *Nat. Med.*, 2000, 6, 1330-1333.

Nagabhushana, K. S.; Ameer, F. and Green, I. R. *Synthetic. Com.* 2001, 719-724.

Nelson, D. L. and Cox, M. M. *Lenhninger Principles of Biochemistry,* third edition, 2002, p. 723724.

Palmer, M. H. *The Structure and Reactions of Heterocyclic Compounds.* London: Edward Arnold, 1967, p.105-144.

Peterson, J. R.; Zjawiony, J. K.; Liu, S.; Hufford, C. D.; Clark, A. M. and Rogers, R. D. *J. Med. Chem. 1992, 35,* 4069-4077.

Ribeiro, C. M. R.; de Souza, P. P.; Ferreira, L. L. D. M.; Pinto, L. A.; de Almeida, L. S. and de Jesus, J. G. *Quim. Nova,* 2008, 31, 759-762.

Riffel, A.; Medina, L. F.; Stefani, V.; Santos, R. C.; Bizani, D. and Brandelli, A. Braz. *J. Med. Biol.*

*Res.,* 2002, 35, 811-818.

Souza, A. E.; Figueiredo D. V.; Esteves, A.; Camara, C. A.; Vargas, M. D.;. Pinto, A. C and Echevarria, A. *Braz. J. of Med. and Biol. Res.,* 2007, 40, 1399-1402.

Sriram, D.; Yogeeswari, P. and Thirumurugan, R., *Bioorg. Med. Chem.* Letters. 2004, 14, 39233924.

Tapia, R. A.; Lizama, C.; López, C. and Valderrama, *J. A. Synthetic. Com.* 2001, 31, 601-606.

Tomson, R. H. in *Naturally Occurring Quinones*; Academic Press, London: New York, 585, 1971.

Tonholo, J.; Freitas, L. R.; de Abeu, F. C.; Azevedo, D. C.; Zani, C. L.; de Oliveira, A. B. and Goulart, M. O. F. *J. Braz. Chem. Soc.* 1998, 9, 163-169.

Tran, T.; Saheba, E.; Arcerio, A. V.; Chavez, V.; Li, Q.; Martinez, L. E. and Primm, T. P. *Bioorg. Med. Chem.* 2004, 12, 4809-4813.

Tsuji, J. John Wiley & Sons Ltd. *Materials Transactions.* England, 1995

Vaitilingam, B.; Nayyar, A.; Palde, P. B.; Monga, V.; Jain, R.; Kaur, S. and Singh, P. P. *Bioorg. Med. Chem.* 2004, 12, 4179-4188.

Waterman, P. G. and Muhammad, I. *Phytochemistry.* 1985, 24, 523-527.

Watson. J. D.;[et al.]. *Molecular Biology of Genes* 5. Ed.- Porto Alegre : Artmed, SA, 2006. 760p.

Wu, J.; Cui, X.; Chen, L.; Jiang, G. and Wu, Y. J. *Am. Chem. Soc.* 2009, 131, 13888-13889. Xu, Z.; Barrow, W. W.; Suling, W. J.; Westbrook, L.; Barrow, E.; Lina, Y. and Flavina, M. T. *Bioorg. Med. Chem.,* 2004, 12, 1199-1207.

Printed by Books on Demand GmbH, Norderstedt / Germany